Working with Your Woodland

Mollie Beattie
Charles Thompson, and
Lynn Levine

WORKING WITH
YOUR WOODLAND

A Landowner's Guide

REVISED EDITION

Illustrations by Nancy Howe
Foreword by Carl Reidel

University Press of New England
Hanover and London

Published by University Press of New England,
One Court Street, Lebanon, NH 03766
www.upne.com

CIP data appear at the end of the book
ISBN–13: 978–0–87451–622–7

University Press of New England is a member of the
Green Press Initiative. The paper used in this book
meets their minimum requirement for recycled paper.

Grateful acknowledgment is made to the publisher
for permission to reprint the excerpt on page 1 from
*A Sand County Almanac, with other essays on
conservation from Round River* by Aldo Leopold.
Copyright © 1949, 1953, 1966, renewed 1977, 1981
by Oxford University Press, Inc.

Printed in the United States of America
10 9 8 7

To the future of the
people of New England and the
forest that surrounds them

Contents

Foreword to the First Edition

Like many New England forest landowners, my reasons for owning a few acres of woodland seem to change with the seasons or who asks. After a new snowfall, or during grouse season, the reasons seem obvious. As the deadline for paying property taxes approach, the old doubts return. And when my forester friends ask about my management plans, I change the subject to the weather or politics. With the exception of my first car—a very used '48 Plymouth—nothing has been a greater source of pleasure and frustration than my 35 acres of Vermont woodland.

I suspect I have thousands of "neighbors" in New England, folks with a few acres or upward of several square miles of forest land—land they love but wish that they understood better and could make a bit more productive. Some would like to realize needed income. Others hope to improve wildlife habitat, preserve scenic values, or ensure a source of fuelwood. For some, goals are unclear because the possibilities are unknown.

Much of the frustration of forest land ownership has come from the difficulty of getting information about the possibilities for management, about sources of professional help, about the costs and benefits of management, or about what to expect if one decides to harvest trees commercially. Many new forest owners without farming experience hesitate to contact public foresters or consultants for fear of revealing ignorance about their own land. "What if I ask a dumb question?" A lot of the frustration of forest ownership disappears once you get started; as it was with my '48 Plymouth, starting is the real challenge.

This is a book about starting. It is a book that has been needed for a long time. Written by three young foresters with first-rate qualifications, it is a practical guide for landowners who want to realize the potential of their forest land. Most important, it is a book based on sound biology and sound economics, which are critical elements for successful forest management.

The authors know their forest science, and weave their expert knowledge into this book in a way that is understandable to nonprofessionals without being superficial. Essential terms are explained, and key concepts are illustrated with functional figures. Unlike some popular books on forestry, Beattie, Thompson, and Levine are not out to sell forestry. They are honest about the initial aesthetic impacts of some management practices, as well as the pitfalls of less-dramatic but ineffective practices like selective harvesting methods. This is an honest, forthright book based on solid scientific forestry; it is equally well-based on sound economics and practical business considerations.

Reliable advice is provided on all aspects of land ownership and forest management, with sections on land evaluation and inventories, contracts for logging and management planning, financial management, and taxes. Suggestions for the "care and feeding" of consulting foresters and loggers will be invaluable for beginners and veterans alike. For those reluctant to get started on managing their forest land, this is a book to build confidence. By decoding forestry jargon and laying out the essential steps in planning and management, the authors seem determined to put landowners in charge of their land. I guess that's what I like best about this book. It respects the reader. It respects the landowner's personal objectives by offering a wide range of alternatives and sound advice on how to make conscious choices from those options.

This is not a do-it-yourself guide. While it looks to the landowner for important decisions about objectives and for choices among management options, it is realistic about an owner's skill and time. You are told how to get technical or legal help, and warned about the cost and pitfalls of homemade remedies. It is a source of practical, sensible advice.

A final word about the book's mood, or outlook. I have long been impatient with the prevailing sense of pessimism among my colleagues in forestry about the future of private woodland management in New England. Problems of land subdivision and owner turnover, coupled with rising taxes and interest rates, are cited endlessly as deterrents to good forest management. Beattie, Thompson, and Levine prefer hope to despair. These imaginative foresters offer new ideas and a challenging array of management options—visions of a future in which demands for energy, wood products, recreation, and other forest values will provide incentives and rewards for land stewardship. Their guide for woodland owners will help us all to reap the many benefits of forest ownership in New England.

University of Vermont   CARL REIDEL
January 1983   *Professor of Forestry*

Foreword to the Revised Edition

It hardly seems possible that I wrote the first Foreword to this book ten years ago! Nor does it seem possible that the Black Walnut seedlings I planted in 1983 are 25 feet tall and yielding nuts.

It is evident in the revisions to this new edition and in the new Postscript, On Stewardship, that the authors have matured as well. They have continued their professional journeys, including leadership of a state forestry agency, presidency of a state foresters' association and ownership of a unique environmental education business. But the past decade has not been easy for people dedicated to sound land management. It would have been easy to have become a pessimist in the face of the local and global threats to our forest that have intensified since 1983.

Despite the somber tone of their new Postscript, it is clear that the authors still "prefer hope to despair," as I said in my 1983 Foreword. They may have become somewhat disillusioned optimists, but they are optimists nevertheless. A bit older and a lot wiser, they have come to see forest stewardship in deeper, more global and ethical contexts. They are even more convinced than a decade ago of the enormous importance of well-managed, healthy forests to the health and future of the planet and all its peoples.

There is very little I would change in my original Foreword, not even the phrase, "written by three young foresters." Despite the seasoning of a decade, these authors still have a lively, young outlook about the future of our forests and the practice of forestry. It is still a book about starting, about planning for the future and taking those first decisive steps to becoming a responsible forest steward. And for those who began their forest management journey after an earlier reading, it remains the essential forestry handbook for the future.

University of Vermont　　　　　　　　　　　　　CARL REIDEL
January 1993　　　　　　　　　　　　　　　　*Professor of Forestry*

Preface to the First Edition

This book is intended as a guide to the sensible use of woodland in New England, to a middle way between overuse of the forest and not using it at all. It is about forest management, and it is meant for the half-million private landowners—those other than forest industries and public agencies—who control more than half of the productive forest land in New England.

In New England, forest management means the manipulation of woodland vegetation to encourage the sustained production of some combination of wood, wildlife, maple sap, or scenery. Timber production is the most familiar goal, but it is only one of many alternatives for forest management.

A forest does not need to be managed. What is "best" for the forest has meaning only in relation to some human use. It is of no benefit to the forest if "poor" trees are removed to encourage the "good" residuals; the quality of trees is a human judgment about their utility for lumber or some other product. Habitat management to encourage game birds is not best for the birds; it is best for the bird hunter.

Nor does a forest need management to produce forest products. Despite two centuries of bankrupt land-use and timber cutting practices, which are still widespread, the New England woods continue to yield lumber, fuel, sap, and game.

Now, however, at the end of the twentieth century, there is mounting concern in New England and the world about the quantity and quality of forest products and amenities: doubts that they can be produced at a rate that will fill an ever-increasing consumer demand. Surely that demand will have to be tempered with conservation. Management of the private forest resource is the only other practical measure that can significantly increase the supply of high-quality forest products in New England, and the capacity of a forest acre to yield a variety of benefits.

We hope this book helps to ensure that those landowners who decide against woodland management do so not because of lack of information on the purposes, processes, economics, and impacts of sound forest practices. Although the decision not to manage is a legitimate one and has advantages, it is best made with an understanding of the benefits foregone, and of the tolerance of the forest ecosystem for human manipulation.

In considering the growing demand for the resources they control, and the intensifying pressure of the cost of living, most forest owners in New England are awakening to the economic values of their woodlands. Our concern in writing this book is that, as the forests are used more intensively, they are used sensibly; that landowners are prepared to make a prudent response to the demands for the resources of their woodlands.

Sensible use of the forest is an acknowledgment that we have become part of the forest's ecology. After two hundred years of extensive logging, farming, and hunting, human activity has rivaled the importance of weather, disease, and fire in shaping the character and composition of the New England woods. Sustainable use of the forest to meet unprecedented demand for its products is only possible through practices that imitate those natural forces. Timber harvesting can mimic the way wind flattens scattered patches of forest, or the way age and disease can select single trees. We can grow trees by accelerating their natural forest life cycles with a series of careful thinnings. Logging erosion, largely the result of poorly constructed woods roads, can be minimized by following the example of wildlife trails, which consistently (though not consciously) follow hillside contours. Wildlife habitat can be improved in quality and quantity if tree harvesting takes the place of forest fires in keeping the forest in a pattern of young and old trees.

Forest landowners will encounter the demand for their forest resources in many ways. For most owners, the increasing values of land and timber may exert pressure to sell, subdivide, or develop land, or to prematurely harvest the merchantable trees. Loggers are now approaching landowners in earnest with offers to purchase timber. Timber management assistance from forest industries is increasingly available as manufacturers of paper and wood products anticipate shortfalls of supply from their own forest lands.

To respond wisely to these overtures, a woodland owner needs some understanding of the life of the forest; a clear definition of the uses he or she wants to make of the land; and an ability to compare the outcomes of various management schemes, to choose one, and to steward its implementation. Beyond equipping owners with these basics, this book does not attempt to help those who have the inclination to become amateur foresters themselves, except to provide references to some good, readable technical information.

Foresters ourselves, we encourage the use of professionals, but only as management consultants hired to apply techniques to implement landowners' choices and to give advice, not as primary decision makers. It is landowners whose vital interests are best served by good forest practices, and whose long-term perspective permits the patience that is a vital element of woodland management. Therefore, they should be the ones to develop and pursue a plan for their share of the New England forest—a resource as valuable and complex as it is beautiful.

Preface to the Revised Edition

Since a central theme of this book is the dynamism of the forest, the book, too, must change. Ten years after its first edition, we have updated it to reflect the responses of New England's forests to significant impacts of weather, disease, insects, and timber harvesting that have occurred over the past decade. Of course, most of the changes in this revised edition reflect developments not in the forest, but in our understanding of it. Our new knowledge and concerns have elevated our standards for good forest management.

We are more concerned now than ten years ago about the ecological impacts of forest management and logging. Protection of wetlands, rare and endangered species, critical wildlife habitats and, in particular, of the ecosystems that support and are supported by each of these are of ever greater importance in forestry. We know more about the negative ecological impacts of forest fragmentation through land subdivision and timber harvesting, and about silvicultural and legal means of avoiding them. Development pressures have reached further into the region's woods, sweeping away important ecosystems and years of forest management. We now believe that good forestry must include legal means of protecting productive lands and ecological values indefinitely into the future.

Many of the revisions to the first edition reflect technological and legal developments. Timber harvesting machinery has itself not changed drastically, but the relative popularity of various types has been significantly affected by energy and economic trends of the last decade. The most important legal changes over the period have been in timber harvesting and environmental protection laws and in the state and federal tax code.

The final chapter of our first edition was an attempt to predict future trends in forestry, most of them technological. Some of these materialized, but most others were preempted by unforeseen events. What was most startling for us was that those critical events were not regional, but global. Our outlook over the last decade has changed with the realization that New England's forests are no longer just New England's. The final chapter of this edition is a postscript on the meaning of that reality.

Working with Your Woodland

I A New England Forest History

We sensed . . . that our saw was biting its way, stroke
by stroke, decade by decade, into the chronology of a
lifetime, written in concentric annual rings of good
oak. It took only a dozen pulls of the saw to transect
the years of our ownership . . .
>   ALDO LEOPOLD, A *Sand County Almanac*

The trees that dominate New England's landscape grow slowly in relation to the span of our own lives, and we see them as permanent and un-
changing, a cherished natural constant in times when change is rapid and
all too apparent. But a forest is anything but static, and the forests of New
England tell two important, intertwining stories of change: one story of the
forests' natural processes, and another of their response to being alternately
used and disregarded by people throughout history. This chapter describes
the interplay between ecological forces and human activities that have
shaped the New England forest, an understanding of which is essential for
anyone approaching the complexities of forest management.

It is a mistake to believe that leaving the forest alone will ensure that it
will stay as it is. The forest changes visibly and dramatically during strong
winds, ice storms, outbreaks of disease and insects, and forest fires. Even between such sudden events, and even without human disturbance, the
forest is dynamic because the woods change slowly but incessantly through
succession. Succession is the constant process by which plants take over an
area, change conditions of soil moisture, of fertility, and especially of light,
then wane as they are replaced by other plant species that find the new
conditions more favorable. (*Forest succession* is the term applied to this
process as it occurs among tree species.) In addition to the changes in soil
fertility and environmental conditions effected by the vegetation on a site,
other factors such as climate, soil type, weather trends, and levels of seed

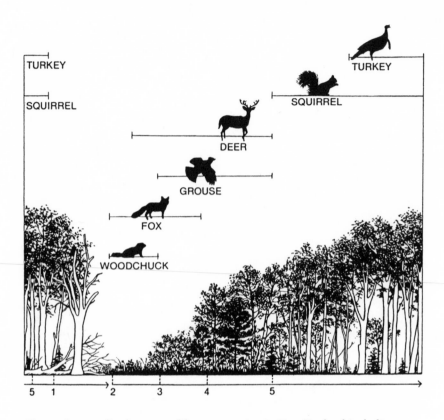

Fig. 1. A generalized pattern of forest succession in New England including animals typical of each successional stage. After a major disturbance in a forest (1), a site may be reclaimed by small plants (2). Then, over perhaps a century or more, the site might be occupied by a series of shrubs (3), sun-loving trees (4), and trees able to regenerate in shade (5). A disturbance could occur at any point in the progression, setting the vegetation back to some earlier stage of succession.

production determine which plants will follow others in succession. Trees do not produce equal amounts of seed each year, and those that produce a good crop when weather conditions are favorable will have a better chance to take hold on a suitable site.

But the factors that direct succession, the interactions among them, and the precise ways in which they effect change in the forest have not all been identified, nor are they fully understood. Elements of chance (such as weather) are also part of the process, so succession is both complex and, to some extent, unpredictable. However, generalized and simplified patterns of succession can be described; an example is given in figure 1. Following a major disturbance, such as extensive windthrow (uprooting of trees by wind), fire, or land clearing, a moist upland area with fertile soil might first be claimed by grasses and brush, then by various sun-loving tree species such as aspen, gray and white birch, cherry, and pines. As their seedlings grow, these species eventually shade the once-cleared area, making it inhospitable for other seedlings of their own species. If the shade is not heavy, shrubs and trees with an ability to withstand some shade, such as oak, ash, and red maple, might form an understory, depending on available seed sources and soil conditions. An *understory* is a layer of small plants, shrubs, or small trees that establishes itself beneath the canopy of the tallest trees, which is the *overstory*. If little sunlight pierces the overstory, only small plants, and tree species able to survive in heavy shade, such as sugar maple, beech, spruce, and hemlock, can establish themselves on the forest floor. As the overstory eventually dies of old age or other causes, the understory will take advantage of the light, moisture, and nutrients previously claimed by the overstory, increase its growth rate, and succeed the old forest. Where there once may have been pine, aspen, cherry, or oak, there is now sugar maple, beech, spruce, or hemlock. If no major disturbance occurs, the forest will eventually be dominated by species that can regenerate in their own shade. Hypothetically, such species could succeed themselves for many generations; however, a disturbance in some form, such as a forest fire, wind, disease, or insect infestation, almost always occurs before this sequence can run its full course.

For instance, hurricanes have battered New England's forests regularly. Figure 2 shows that severe storms have occurred frequently in much of the region. Table 1 lists the major hurricanes that have struck various parts of the region over a 320-year period.

It is important to remember that the succession pattern illustrated in figure 1 is a generalization of one path that succession might follow on a particular type of site. When the last glacier retreated from New England around ten thousand years ago, it left highly varied terrain and soils. Since plant and tree species are each adapted to particular soils, sites, and environmental conditions, the pattern of succession will vary from site to site.

Long before humans dwelt in New England, the dynamics of succession were operating; the typical succession was from small plants to shrubs, to sun-loving trees, to species able to survive in the understory. At some point a catastrophe would interrupt the process. During less violent disturbances

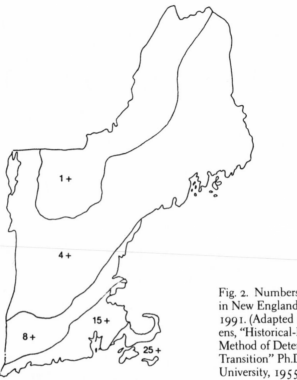

1 +

4 +

15 +

8 +

25 +

Fig. 2. Numbers of hurricanes in New England from 1635 to 1991. (Adapted from E. P. Stevens, "Historical-Developmental Method of Determining Forest Transition" Ph.D. diss., Harvard University, 1955).

small gaps were created in the overstory, and where sunlight was let in to the forest floor the understory was released or the less shade-tolerant species were invited to gain a foothold. During the most extreme disturbances large areas of the forest were burned up or blown down, returning the site to some early stage of succession: sun-loving trees, shrubs, or grasses.

All parts of a biological system are connected; as forest vegetation changes through the process of succession, other members of the forest community are affected. Because different vegetation provides food and cover for different animals, alterations in the plant and tree composition of the forest mean that wildlife habitat changes too. If a successional stage that includes a major source of food or shelter for an animal disappears or flourishes, that species' population will tend to dwindle or increase, respectively (see fig. 1). For instance, deer depend largely on young forest and the shrubby edges of open areas for food; to the extent that such areas are unavailable, the deer population will be limited.

While change through natural disturbance and succession is one thread permeating the history of our forests, direct human influence is the second. The relatively brief history of land use in New England, summarized in figure 3, shows that the appearance, extent, and even the species composition of today's forests have been affected by human use. Before the arrival of European settlers, woodland Indians cleared land near rivers and major inland lakes for food production. Because the Indian population was never large, most of the New England forest was unaffected by these practices. Except for wetlands, a few barren mountaintops, and major river valleys cleared by the Indians, the primeval forest dominated the region. Elk, caribou, mountain lion, and timber wolves populated this forest. Animals such as deer, grouse, hare, skunk, and quail were largely confined to the cleared land near Indian settlements and to early successional forests that had resulted from natural disturbances.

When European settlement of the region began in the mid-1600s wood was the household fuel. Near settlements, overcutting of the forests for fuel and lumber began early, and by 1700 some communities had enacted ordinances to regulate cutting.

Fuelwood harvesting has been significant and continuous from the time of colonial settlement until the present day, varying in this century with

Table 1 Major Hurricanes in New England, 1635–1991

Year	Area	Intensity
1635	Massachusetts, Connecticut, Rhode Island	Extreme
1638	Massachusetts, Connecticut, Rhode Island	Extreme
1723	Southeastern New England	Major
1761	Southeastern coastal New England	Major
1815	All areas	Extreme
1821	Southern New England	Major
1866	Southeastern coastal New England	Major
1869	Southern New England	Major
1938	All areas	Extreme
1944	Coastal areas	Extreme
1954	Coastal areas	Extreme
1955	Southern New England	Extreme
1960	Coastal areas	Major
1985	Southern coastal New England	Major
1991	Coastal areas	Extreme

Sources: G. E. Dunn and B. I. Miller, *Atlantic Hurricanes* (Baton Rouge, La.: Louisiana State University Press) © 1960 and 1964. Reprinted by permission; Harvard Forest, Petersham, MA.

Fig. 3. Summary of New England forest history from 1600 to present, by decade.

1600	Forest 96 percent of land area
1620	European settlement begins
	Subsistence farming begins
1630	First sawmill
1700	First local regulation of overcutting near Boston
	Deer herd small
1750	Elk, caribou virtually eliminated
1770	American Revolution (1776)
	Forest 91 percent of land area
1790	International lumber markets expand
1800	Transition from subsistence to market farming complete
	Great increases in populations of open-land wildlife
1820	Erie Canal opens (1825)
	Sheep industry gains in significance
	Deer rare in much of region
1830	Beginning of farm abandonment
1840	Extensive railroad expansion
1850	Gold Rush (1849)
	Arrival of the Eastern cottontail; wild turkey eliminated; beaver rare
1860	Civil War Homestead Act (1862)
1870	First pulpmill; intensive harvesting for charcoal manufacture
	Moose eliminated from much of its range
	Gypsy moth introduced
1880	Oil and gas gain significance as heating fuels
	Peak of cleared land
	Peak of annual fuelwood use at 4.1 million cords
	Forest 58 percent of land area; lowest historical extent

1890	Sheep industry withers; land abandonment intensifies
	Gypsy moth outbreak
	End of regional lumber self-sufficiency
	Mountain lion eliminated or rare
1900	"Boxboard boom": peak of annual lumber production
	Wolf eliminated
	Introduction of Dutch elm disease, chestnut blight, white pine blister rust
1910	Spruce budworm outbreak; beech bark disease introduced
1920	Invention of waxed papers that replace wooden boxes
	Significant purchase of land for recreation/second homes begins
	Peak of deer population; resumptions of legal hunting
	Gypsy moth outbreak
	First regional inventory shows timber harvest far exceeds growth
1930	Great Depression (1929–1939); hurricane of extreme intensity (1938)
1940	Beaver reestablished; coyote arrives
	Gypsy moth outbreak; chestnut virtually eliminated
	Forest 74 percent of land area
1950	Tremendous intensification of recreational interest in land
	Annual pulpwood harvest surpasses both the annual fuelwood and lumber harvests
	Gypsy moth outbreak
	Estimated 40 percent of the forest less than 20 years old
1970	Oil embargo (1973)
	Annual fuelwood use ½ million cords
	Restoration of the wild turkey; moose moves back southward
	Spruce budworm outbreak
1980	Renewed interest in New England timber; expansion of international markets
	Annual fuelwood use of 3½ million cords approaches both the annual pulpwood and lumber harvest and the historical peak of fuelwood use
	Gypsy moth outbreak
	Forest 80 percent of land area; regional inventory shows growth twice the annual harvest
1990	Lowered oil prices; decline of fuelwood use
	Forest health a major concern; outbreaks of new insects and diseases
	Ongoing recreational and suburban subdivision of forests

the price of oil. Another practice that has been maintained from the 1600s to the present is *highgrading*—taking only the best trees in a stand. This practice, followed for three centuries, has had a profound effect on the New England forest: each harvest of the biggest and best-formed trees further lowered the quality of the forest that remained. Even in remote areas of New England highgrading started early. The best trees—especially white pines—were cut, and floated down rivers to sawmills. England had already overcut its own forests, and the great timber wealth of New England proved to be strategically important. England was able to maintain naval superiority with the help of oak timbers and pine masts originating in the New England forests. The Broad Arrow Acts of 1691 and later ruled that all pines greater than twenty-five inches in diameter were the king's property, to be reserved as masts for the Royal Navy. These laws were a major source of irritation to the colonists, who ignored the edicts and even set a few fires to deny the king his trees.

Elk and caribou, true wilderness animals, disappeared from New England as forest clearing began. Market hunting soon depleted wild turkey populations, and the beaver was rapidly extirpated from southern New England when its pelt became a medium of exchange in the colonies. As the beaver dwindled, so did the otter, mink, and muskrat for which beaver ponds are an important habitat element.

The heavy cutting of overstory trees allowed more sunlight to reach the ground, stimulating the growth of shrubs, herbaceous plants, sunloving trees, and sprouts from stumps. The result was a sudden change in wildlife habitat conditions. Species preferring the conditions offered by mature forests suffered, while animals like deer, which need brushy field-forest edges, flourished. In turn, the deer (and the settlers' livestock) were food for the timber wolf, which multiplied far beyond its presettlement numbers and became the subject of intensive eradication attempts funded with large percentages of local taxes.

Following the American Revolution, land clearing in all but the remote and difficult areas accelerated as subsistence farms grew to commercial size to feed growing manufacturing centers to the south and east. The New England farmers toiled to remove the annual crop of stones from the land, depositing them in thousands of miles of walls that twentieth-century New Englanders cherish (frontispiece). In addition, increasing amounts of wood were needed as raw material for industry, and to provide charcoal for manufacturing processes such as iron smelting. Parts of the Massachusetts and

Connecticut forest were cut repeatedly to supply charcoal and fuel from oak and chestnut sprouts that grew vigorously from stumps (fig. 49). This legacy of evenaged sprout hardwoods is still visible in today's woods. Potash and pearl ash for fertilizer and glassmaking were produced by piling and burning unmerchantable timber during land clearing. When steam engines appeared, large sections of Vermont and New Hampshire forests were cut to fuel their boilers. At the end of the 1700s, international markets for New England lumber had been opened, adding to the drain on the forest. By 1840, much of the New England landscape was like a photographic negative of itself before settlement: not a thick forest punctuated by small openings, but a shorn landscape with scattered tufts of trees. Sixteen million acres of forest had been cut to run factories, make steam, charcoal, and potash, or simply to clear the land for farming and a growing sheep industry. In 1880, an estimated 42 percent of New England was open, with significant forest cover persisting only in the region's high elevations and northern reaches.

Gone was the forested habitat of the wild turkey, which depended on large blocks of undisturbed hardwoods. The moose, portions of whose range in the spruce and fir forests had been decimated, was pushed from all but the northernmost reaches of the region. Although young forest and brushy forest edges are important food sources for deer, these animals also depend on some tall forest for cover, and especially on softwoods for winter shelter. Lacking cover and under heavy pressure from hunters, deer were pushed from many parts of New England by the mid-1800s. At the same time, the loss for woodland wildlife was a gain for animals of open land. Skunk and woodchuck populations exploded, and, by mid-century, the gray fox and cottontail rabbit had extended their natural ranges northward from the mid-Atlantic region into the now-open landscape of New England.

Continuing industrialization, the opening of the Erie Canal in 1825, the California gold rush in mid-century, the advent of the railroads, and offers of free western land to Civil War veterans began a long, steady agricultural decline and a period of large-scale land abandonment in most of New England. The higher wages and fancy life-style of the cities, tales of mining fortunes to be made, and the fertility of western farmland proved irresistible to those who had struggled with the rocky, thin soils on the hill farms of New England. The farmers went west; grains, meat, and wool came east on the railroads and through the Erie Canal to be sold at prices too low for competition from New England farms. Abandoned roads, cellar

Fig. 4. White pine claiming an abandoned pasture.

holes, lilac bushes, apple trees, and sometimes a potentially dangerous shallow well are all that remain of some of these farms. Deed descriptions often became vague and unclear when land changed hands during the remainder of the 1800s, and when many of the competent land surveyors followed the farmers west. Prospective buyers of woodland today are often presented with poor property descriptions which have been perpetuated since the time of farm abandonment.

Large-scale land abandonment restored the dominance of natural processes in the New England woods. Forests quickly took over the abandoned land, eventually regaining eight of the sixteen million acres cleared. Depending on the location within the region, seeds of pine, spruce, red cedar, or fir, which were able to penetrate the sod and take root in the soil beneath, established almost pure conifer stands in fields and pastures (fig. 4). The trees gradually formed a closed canopy, stifling the grasses and shrubs by cutting off their sunlight.

Much wildlife rebounded with the forest. The most notable comeback was that of the deer, although this was not always accomplished without the assistance of restocking. In 1878, seventeen deer were brought from New York State to Vermont to restore the species; today Vermont supports one of the highest deer densities in New England.

Expansive pine stands growing in full sunlight on abandoned farmland encouraged a native insect, the white pine weevil. This insect feeds on the top, upright branch, or "leader," of white pine (and sometimes red spruce), killing the branch and forcing side branches to take over as substitute leaders; this results in tree deformity. Since they are especially active in stands where trees are all of the same size and thus have their crowns in full sun, populations of the sunloving weevil were rampant in the young pine stands in abandoned pastures. Deformed "cabbage" pines are visible all over New England (fig. 5).

Gradually conditions on the woodland floor beneath the new forests of pine, cedar, fir, and spruce inhibited the growth of seedlings of these species, and hardwood species better adapted to shade were able to take hold and survive.

By 1900, the pine and spruce on the earliest-abandoned farmland had grown to merchantable size. Human impact on the composition of the forest again became pronounced. Timber was removed in great quantities, this time largely for wooden shipping boxes or spruce lumber. In 1909, the "boxboard boom" pushed New England lumber production to its historical peak of three billion board feet, three times the region's current annual production. In 1920, in the first official assessment of New England's timber resource, it was estimated that the annual "drain"—the amount of wood being cut in the region each year—was half again as much as what was being grown. Sawdust piles from that era may be still visible on some woodlots.

When the big pine and spruce were gone, the long-suppressed hardwoods that had seeded beneath them were flooded with sunlight. The deer again gained. Although their winter cover was reduced as the softwoods fell, the loss was not as significant as it had been during land clearing for agriculture a century before. The succeeding hardwoods provided abundant browse, and the deer population swelled. In 1910, a century-long deer hunting ban was lifted in Massachusetts.

Of even more significant and lasting human impact than harvesting was the introduction of several forest insects and diseases to New England, mostly in the early 1900s. After a fungus was imported from Asia on nursery stock in about 1900, the American chestnut was virtually eliminated as a timber tree within fifteen years in New England, and within thirty years in North America. The chestnut was one of the most common trees in central and southern New England, and was prized for its beauty,

Fig. 5. "Cabbage" pines. When its topmost, upright branch was killed by the white pine weevil, the young white pine in the foreground responded with vertical growth of its side branches, giving the tree a spherical shape; hence the nickname "cabbage" pine. The tree in the background, weeviled as a young pine standing in an abandoned pasture at the turn of the century, shows the multistemmed form characteristic of mature cabbage pines.

durable lumber, and nuts, which were a human food and prime forage for many wildlife species. Before its demise, which was complete by the 1930s, chestnut had constituted one-third of all potential sawtimber in the Connecticut woods. In 1904 Dutch elm disease was discovered in Boston, probably having been transported to the United States from Europe in lumber. The disease, which is fatal to the American elm, has spread throughout the tree's range. Research on these diseases continues today, and there is some hope that they will eventually be defeated.

Defoliation by the gypsy moth, introduced to the Boston area from France in the mid-nineteenth century, reached threatening proportions in the early 1900s. The insect has since spread throughout New England and into the Middle Atlantic states, periodically ravaging millions of acres of trees. People continue to spread the insect, whose sticky buff-colored egg masses often catch a ride on a car going south or west for the winter. Outbreaks of the gypsy moth are cyclical, often beginning in oak stands on dry sites. Many such stands occupy sites which were once dominated by pine and were heavily logged and sometimes burned; the oaks were well-suited to taking over these severely disturbed areas.

White pine blister rust, a fungal disease which severely depreciates wood quality and which is eventually fatal to a tree, was also introduced to the United States in a shipment of infected lumber at the turn of the century. Success in efforts to reduce gooseberry and currant bushes on which the fungus relies for part of its life cycle, or perhaps a decrease in the extent of pine stands seems to have diminished the danger of a widespread blister rust problem.

Another example of imported pests and diseases that have had a long-lasting, perhaps permanent, effect on New England's forests is a European insect that facilitates the spread of beech bark disease, and was first documented in Nova Scotia in 1920. The disease is now causing widespread mortality among beeches throughout the tree's range in the United States. Many of these diseases and insect pests have had severe impacts on wildlife populations. Elimination of the chestnut and extensive beech mortality represent major losses of forage for bear, deer, turkey, squirrel, and other wildlife species.

The pace of introduction of exotic pests seems not to have flagged. A severe Asian threat to sugar maple, the pear thrips, was detected in New England in the late 1980s. Hemlock may be decimated by the Hemlock wooly adelgid, a recently introduced pest but of unknown origin.

Fig. 6. A Civilian Conservation Corps plantation. The CCC planted much spruce and pine on abandoned farmland throughout New England during the Great Depression. The plantations, now about 60 years old, are still common in the New England landscape.

After the huge timber harvests in the early years of the twentieth century, the annual harvest of lumber declined steadily until the 1938 hurricane and World War II. Fuelwood use peaked at 4.1 million cords in 1880, but soon began to decline as coal and oil became popular. Where the supply of pine held out, the pine box and cooperage industry survived into the first decades of the 1900s, until the invention of waxed papers and other alternative packaging. Otherwise, demand for New England's forest products decreased steadily between the two world wars, although drain still slightly exceeded growth. Other regions of the country with reserves of old-growth timber could supply New England's needs more efficiently than could her native woodland, in part because of cheaper transportation, especially after the opening of the Panama Canal in 1919. Harvesting in most of southern and central New England was now reduced to its historical minimum—the cutting of fuelwood and the best sawlogs. In the remote northern areas of New Hampshire, Vermont, and Maine, the supply of large pine and spruce was exhausted by 1900. Harvesting did increase in northern New England until the 1930s, but for paper, not lumber. Processes for pulping wood had been developed in the late 1800s, and the long fibers of spruce were ideally suited to papermaking. Clearcutting in the northern spruce/fir forests was gaining momentum in the early 1900s, and pulpwood became a new merchantable forest product. The use of hardwoods for pulp began in 1912.

During the depression forest industries stagnated. Cutting for timber and pulp continued, but at declining rates. The woods, like the fields a century before, were quiet except for fuelwood harvesting and the activities of the Civilian Conservation Corps workers, who built recreational facilities and fire roads, thinned stands of trees, and planted spruce and pine on abandoned farmland (fig. 6).

The major hurricane of 1938 blew down three billion board feet of timber in southern and central New England, an amount equivalent to the region's peak annual lumber production two decades earlier. An entire government program, the New England Timber Salvage Administration, was created to buy the salvaged timber, which was dumped into ponds for protection from insect attack and decay. This huge inventory of logs supplied nearby mills until World War II, when the remaining salvage was requisitioned for the war effort.

After World War II timber from private land was a minor component of New England's economy. Construction lumber continued to come from the far west and Canada, where there were reserves of old, high-quality timber. Eventually lumber also came to New England from the southeastern states, from flat pine lands where timber is grown in a third of the time it takes to grow in New England. Cheap and abundant oil trivialized wood as a fuel, and the region's consumption fell to half a million cords in the 1960s.

It was competition from other regions that depressed the demand for New England timber during this period, but a contributing factor was the large portion of the region's woods that was too young to provide sawtimber. Those areas clearcut for pine in the first twenty years of the century were now stocked with small hardwoods. Forests in the paths of the 1938 hurricane, the Maine fire that burned two hundred thousand acres in 1947, and the spruce budworm outbreak of 1912 – 1920 were still recovering. In the 1940s it was estimated that about 40 percent of New England's forest was less than twenty years old.

Timber quality was declining. Many of the region's older stands, which had been left unscathed by natural catastrophes, had by mid-century been highgraded repeatedly for as long as two hundred years, and contained increasing proportions of trees too poor for sawtimber.

One other development in New England forest history, which was perhaps both a result and a cause of the diminishing importance of New England's timber resource during much of this century, has been the purchase of large amounts of land by city dwellers for recreation and speculative investment. This trend became noticeable in the 1920s, gained momentum after World War II, and was rampant by the 1960s. Its impact on New England's forest resource may prove to be as significant as any other event to date. Not only has timber production been superceded by other motives for owning land, but the clamor for country real estate has greatly contributed to the ongoing subdivision (or "parcelization") of land, to the extent that the median size of forest ownerships is now about 10 acres in much of New England—a size too small to be efficiently managed for timber production.

Through the second half of the twentieth century, the volume growth of timber in New England has far outpaced the harvest, a reversal from the previous decades of overcutting. Between 1952 and 1976, for instance, a regional inventory showed the volume growth of timber on private lands to be 56 percent, while the amount cut increased by only 2 percent.

As the forest has reclaimed New England, the region's deer herd has continued to prosper, although its size has shrunk somewhat as the extensive early successional forests of the 1920s and 1930s have matured. The wolf has been long gone from the region, but it has been gradually replaced in some areas by the eastern coyote, which has extended its range into New England from the Great Lakes region. Once again beaver and wild turkey thrive in the forest, after being restored to the region by wildlife management agencies. Moose populations have rebounded in Maine, and have edged southward from Maine and Canada into New Hampshire and Vermont. Several decades of unconfirmed sightings as far south as Massachusetts hint at the recoupment of a small population of mountain lion.

In a cycle of use, abandonment, regrowth, exploitation, and neglect, New England's privately owned woodlands were in the quiet part of the pattern for most of this century. In recent decades, the distant view in New England has not changed greatly: a survey in 1976 estimated the region's annual forest growth to be twice that of the drain from harvests. However, as the century closes, there is again increasing activity due to the familiar economic causes. But, this time, the pattern is complicated by the threats to the health of the forest from new pests and diseases and from atmospheric and stratospheric pollution such as "acid rain" and the "greenhouse effect."

The legacy left to forest landowners in New England is forest soils that were once heavily grazed and cultivated; habitat for those animals able to persist through three centuries of changing and sometimes difficult environments; those trees passed over as too poor to be useful and spared the ravages of imported insects and disease; and neglected stands of young timber too overcrowded to grow well. Hardly pristine, our woodlands are living examples of nature modified by human use and abuse. But they are also a testimony to the incredible resilience of the New England forest. Whether the land has been logged, burned, or plowed, the forest responds with a green cover to stabilize the soil, generate oxygen, filter noise and pollutants, moderate temperatures, and provide shelter and food for animals, and beauty and wood products for people.

The two intertwining stories summarized in this chapter—that of the way in which the forest changes naturally through succession, and that of how it has been changed by more than three hundred years of intensive human use—provide insights that are the first steps toward forest management. When the woods are perceived as static, it is difficult to see how they can be guided and regulated. To understand forest succession and natural disturbance is to realize that it is possible to imitate those processes through forestry practices, and to use them for the production of wood, wildlife habitat, and other benefits.

And there is another conclusion to be drawn from the story of three hundred years of human use: that it is idle speculation to concern ourselves with what the New England forest might be in the absence of the logger, the farmer, the hunter, the sugarmaker, or even the hiker. Human use is now so widespread, intense, and frequent in the forests of New England that the only realistic way to think of ourselves is as part of the forest environment. For landowners who intend to use their woods, an understanding of that relationship and its responsibilities is essential to sustainable forest management.

2　Assessing Woodland Potential

Typically, a person's initial attraction to a piece of woodland has little to do with its prospect for forest management. Yet, eventually, most owners become interested in timber, maple products, or fuelwood, many in wildlife habitat management and the development of trails and vistas.

To decide among management options, a landowner must evaluate a tract's potential for certain products or uses. This evaluation process involves making a checklist of the land and forest features that are important for those products and uses. The list might include the size of a property, its accessibility, the nature of its soils, and vegetational features. The list of factors and their relative importance will vary for each woodland use. For example, some of the factors important for timber production are not important for wildlife habitat improvement.

Timber

Timber management is a continuous process of cultivating, harvesting, and regenerating trees having the best potential for conversion to valuable wood products, such as lumber and veneer. Evaluating woodland for its potential response to timber management involves consideration of several factors: tract size, accessibility, terrain, soils, and the species, size, age, quality, and crowdedness of trees.

The larger a forest ownership, the more frequent are the opportunities for income. In New England, on average pieces of woodland less than 25 acres in size, economically efficient timber production may require intensive management. Even then, smaller tracts can provide timber sales only at long intervals, perhaps up to several decades. A few hundred acres of New England woodland are usually necessary for timber to be a source of income every five years or less.

"Economically efficient" management means that the costs of labor, equipment, and technical assistance necessary to cultivate and protect sawtimber can at least be recouped through timber sales. This becomes increasingly difficult as acreage decreases because the per-acre fixed costs of land ownership, equipment purchase and operation, private forestry assistance, and record keeping may become uneconomically high. Furthermore, the total amount of valuable wood initially available on small tracts of previously unmanaged woodland is likely to be small.

Timber is saleable when it can yield a reasonable profit to a buyer, usually a logger or sawmill. The minimum amount of timber necessary for a viable sale is related to the quality of the timber. It may take a good deal of firewood, but a much smaller volume of high-quality logs, to interest a logger in transporting and setting up equipment, building roads into a woodlot, and hauling wood some distance for delivery.

In timber production, good access and favorable potential for tree growth on a woodland site will compensate for small acreage. Good sites are capable of producing two or three times as much wood per year as poor sites. However, the smaller the tract, the more singular timber management must be; management for such other woodland uses as wildlife habitat improvement, sugarbush development, and perhaps even aesthetic enjoyment might become subordinate if the owner's dedication to timber production is absolute.

All six New England states offer reduced property taxes on qualifying forest land. These tax programs, most of which require a minimum acreage and some level of timber management, are listed and described in appendix 3.

Woodland terrain is very significant in timber production. Forest management for timber (and for most uses) involves removing trees, and as such is restricted to areas where logging equipment can operate. The terrain feature most limiting to equipment is steepness; steepness is commonly expressed as a percentage calculated by dividing the horizontal distance covered by a slope (the "run") into the elevation gained (the "rise"), as shown in figure 7. The types of logging equipment that are currently conventional cannot operate safely or economically on slopes greater than 50 percent. Farm tractors are safe only on level ground, and horses have difficulty dragging, or *skidding*, logs uphill.

Regardless of the mechanical limitations imposed by excessive slope, good forest practice often dictates more severe constraints. On slopes greater

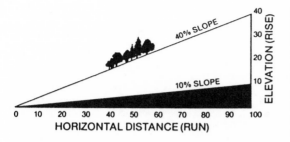

Fig. 7. Slope percent. Slope percent is calculated by dividing the rise, or elevation, by the run, or horizontal distance. For example, a slope that gains 10 feet of elevation over 100 feet of horizontal distance is a 10 percent slope: 10 ÷ 100 = 10%.

than 30 percent, extensive and often costly precautions must be taken to avoid the soil erosion that may result from skidding logs and the building of logging roads. The slope, or grade, of logging roads should rarely exceed 10 percent, and ideally should be kept to 5 or 6 percent.

In general, then, woodland which is on or is accessible from slopes of more than 30 percent is workable only with a great deal of caution, and usually with some additional expense.

In addition to presenting limitations to logging, terrain may also indicate the timber productivity of a site. Frequent rock outcroppings often indicate thin soil and low moisture, and consequently tree growth that is too slow to warrant intensive timber management. Uneven terrain, composed of high spots and depressions called *pit-and-mound relief*, is frequently a sign of poor drainage or thin soil: the depressions are cavities left by blown-over trees which were unable to root securely. Figure 8 illustrates the process that results in pit-and-mound relief. In both thin and wet soils uprooted trees are common and tree growth is likely to be poor. *Aspect*, or the direction in which a slope faces, affects the productivity of a stand of trees, as does the stand's position on the slope. The most productive timber sites in New England are often on lower, northeastern-facing slopes, because they are moist and usually have deep soil. Southern slopes, which are the warmest and driest, and upper slopes and ridges, which have thin soil, can be poor sites for growing timber.

Physical obstacles such as excessively steep slopes, watercourses, extensive swamps, surface boulders, and large rock outcroppings can significantly diminish, or even cancel, the economic value of timber that lies beyond them. They affect economic value by inflating road-building costs, and especially timber removal costs, beyond the sale price of the timber. Deep, wide watercourses particularly may add costs to a logging job, since

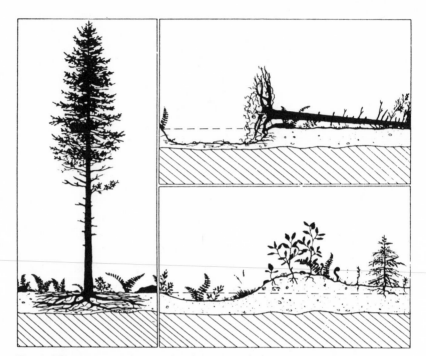

Fig. 8. The sequence of events that forms a pit and mound. On sites that have a high water table or that are shallow to bedrock, trees are prevented from establishing a deep root system (*left*), and become increasingly vulnerable to windthrow as they increase in height. If the anchoring ability of its roots fails and a tree is uprooted, a large ball of soil is pulled up by the root system, leaving a pit (*top right*). As the tree decays, the soil and rotting root system are deposited in a mound (*bottom right*). An area in which the terrain is made up of pits and mounds may be very poor for growing timber.

considerations of water quality often require the installation of culverts or bridges. When a long road is necessary to gain access to a stand of trees, harvesting the timber may not be economical. Finally, trees which can be skidded downhill to a road are, in effect, more valuable than those of equal quality which must be skidded uphill.

It is true that where financial gain is not the primary purpose of ownership, the costs of roadbuilding might not be charged against the value of the forest products at the end of the road; instead, these costs might be regarded as the price paid for improved access to the land. In such cases, the physical problems of access are serious only if they are so extreme as to make loggers unwilling to work on a site.

The cost of moving trees from the stump to the roadside is the most significant financial factor in a logging job; it is determined in part by the length of the haul and the amount of wood that can be transported in a single load, or *hitch*. Most logging jobs involve skidding trees along the ground; the effects of distance and load size on skidding costs are illustrated in table 2. The dollar figures in the table are exemplary only, and should not be taken as absolute since labor and equipment costs are subject to change. On steep slopes and rough terrain, where a machine must haul less than its capacity, the value of a cord of wood or a thousand board feet decreases. Small loads are expensive. Most loggers agree that 600 to 800 feet is an economic average skidding distance for medium-sized skidding equipment (crawler tractors, for example), and that 1,300 to 2,000 feet is an economic distance for large machinery (rubber-tired skidders), although tree quantity, quality, and size will greatly affect these limits. A single horse can handle a 350-foot skid; a team of two horses, a 700-foot skid.

Aside from physical obstructions, difficulties of access may include legal obstacles. Landlocked tracts—those without frontage on a public road—

Table 2 An Example of the Relation of Skidding Costs
to Distance and Volume

Skidding Distance (feet)	Board Feet Skidded per Trip	Cost per Thousand Board Feet
500	400	$23.75
500	600	15.91
500	800	11.96
500	1,000	9.59
1,000	400	28.41
1,000	600	19.07
1,000	800	14.39
1,000	1,000	11.56
1,500	400	33.22
1,500	600	22.38
1,500	800	16.91
1,500	1,000	13.60
2,500	400	43.14
2,500	600	29.14
2,500	800	22.07
2,500	1,000	17.80

Adapted from: J. L. Kroger, *Factors Affecting the Production of Rubber-Tired Skidders*, Technical Note B-18 (Norris, Tenn.: Tennessee Valley Authority, 1976), p. 45.

may present problems that are best solved by a legally deeded right-of-way through neighboring ownerships. Even in cases where one owner has historically used the roads or trails of another to gain access to a parcel of woodland, a precisely worded agreement should be written and signed if management of the woodland will require increased use or development of the roads. In some New England states, laws provide procedures for obtaining a temporary right-of-way to landlocked parcels (see appendix 3 for specific state information).

In examining a piece of woodland, it is advisable to look for old woods roads, which may provide a head start in constructing access to the property. Existing roads should be carefully evaluated. In some cases, it may make sense to design a new road system. Old roads that were adequate for use by draft animals and small logging equipment may not be adequate for heavy, modern equipment.

Soil characteristics are very important in determining the biological potential of a site for regenerating, growing, and harvesting timber. There are definable types of soils, distinguished primarily by layers of different texture, structure, and color that are evident when a pit is dug. Soil types are further defined by their slope and stoniness. The Soil Conservation Service (SCS) of the U.S. Department of Agriculture rates each soil type according to various forest management criteria. The criteria of most interest to landowners considering timber management would be: (1) the productivity of the soil, often expressed as the estimated number of cubic feet of wood which the soil can grow per acre per year; (2) the soil's erodibility; (3) the limitations the soil offers for the operation of conventional logging equipment; (4) the probability of seedling survival; (5) the probability of *windthrow*; and (6) the limitations imposed by the soil on the construction of woodland roads.

Although the fertility and moisture-holding capacity of a forest soil, as measured by its estimated wood production capacity, can be drastically reduced by poor land use, there are currently no practical, economically feasible means of improving soil fertility. The high costs of chemical fertilization, which is practiced in the fast-growing southern forests, cannot be justified economically in the naturally slower-growing forests of New England. Fertilization trials have had erratic results in New England.

Erodibility (erosion hazard) refers to the risk (low, medium, or high) of having wind and water transport significant amounts of soil from a site after normal disturbances caused by the common woodland management practices of timber harvesting. The level of erosion hazard therefore indirectly indicates whether a substantial investment will be necessary for erosion control during timber harvesting. Soils on steep slopes have the highest erodibility.

Equipment limitations may be necessary because of soil and terrain characteristics such as wetness, large boulders, or steep slopes that restrict or prohibit the use of conventional machinery for logging, road construction, or fire control. Predictions of seedling survival take into account the anticipated loss of seedlings (seedling mortality) that occurs especially as a result of unfavorable soil characteristics such as dryness or frost heaving, or of topographic features which promote frost or excessive temperature. Even if healthy seedlings germinate in adequate numbers, survival will be low if soil conditions are unfavorable. Windthrow hazard ratings are based on an evaluation of soil characteristics (especially wetness and the depth to bedrock) that control the development of tree roots and thus affect the ability of trees to stand against the wind. Road construction limitations are rated as slight, moderate, or severe; poor drainage, rock outcroppings, and rocky soils are the most common causes of a severe rating. On such difficult sites, substantial investment may be necessary before timber harvesting can take place.

The SCS maps the soil types of each New England county and identifies the suitability of each soil for timber and wildlife habitat production. For some counties, this information is available in "Soil Survey" reports published by the SCS. Addresses of SCS offices are listed in appendix 4. When using an SCS soil survey, it is important to understand that the smallest unit mapped may be several acres in size, and that on the maps broad areas of a single soil type may actually include up to 35 percent of other soil types. Soils on slopes and other typically wooded areas may not have been mapped as carefully as lands suitable for farming and development. For this reason, when assessing woodland potential, soil survey information should be used in conjunction with ground observations of terrain and on-site measurements of productivity.

SCS agents are available for visits to a property at no charge to the landowner. An agent can prepare a short assessment of the property's potential for various uses through evaluations of soils provided on "Soil Survey Interpretation" sheets, as shown in figure 9. Such soil descriptions provide a preliminary understanding of a soil's potential and its limitations. It should be understood that soil limitations are not strictly insurmountable; with some investment, most can be overcome.

It is imperative that an owner be able to identify accurately the tree spe-

SOIL INTERPRETATIONS RECORD

MA0104

MLRA(S): 143, 144A, 144B
REV. RM. 11-89
TYPIC HAPLORTHODS, COARSE-LOAMY, MIXED, FRIGID

Berkshire series consists of very deep, well-drained soils on uplands. They formed in glacial till. Typically, these soils have a dark brown, rubbly, fine sandy loam surface, 6 inches thick, over 2 inches of light gray fine sandy loam. The subsoil layers, from 8 to 22 inches, are dark reddish-brown, yellowish-red, and yellowish-brown fine sandy loam. The substratum, from 22 to 65 inches, is light olive-brown fine sandy loam. Slopes range from 3 to 75 percent.

BERKSHIRE SERIES RUBBLY

MA0104

RECREATIONAL DEVELOPMENT (C)

Camp areas	3-15%: Severe-large stones 15+%: Severe-slope, large stones	Playgrounds	3-6%: Severe-large stones 6+%: Severe-slope, large stones
Picnic areas	3-15%: Severe-large stones 15+%: Severe-slope, large stones	Paths and Trails	3-25%: Severe-large stones 25+%: Severe-large stones, slope

REGIONAL INTERPRETATIONS

WOODLAND SUITABILITY (C)

Class-determining phase	ORD SYM	Management Problems					Potential Productivity			
		Eros'n hazard	Equip. limit	Seedl. mort'y	Windth hazard	Plant compet	Common Trees	Site Index	Prod class	Trees to Plant
3-35% RB	9x	slight	moder.	slight	slight		Eastern white pine	72 *	9	Eastern white pine
35+%	9R	moder.	severe	slight	slight		Sugar maple	52 *	2	Red pine
							Red spruce	50 *	8	White spruce
							White ash	62 *	3	Balsam fir
							Yellow birch	55	2	
							Paper birch	60	4	
							Balsam fir	60	8	
							White spruce	55	9	
							Red pine	65	8	

WILDLIFE HABITAT SUITABILITY (D)

Class-determining phase	Potential for Habitat Elements								Potential as Habitat for:			
	Grain & seed	Grass & legume	Wild herb.	Harded trees	Conifer plants	Shrubs	Wetland plants	Shallow water	Openld wildlf	Woodld wildlf	Wetland wildlf	Rangeld wildlf
All	v. poor	v. poor	v. poor	good	good	—	v. poor	v. poor	v. poor	fair	v. poor	—

Footnotes

C Ratings based on national forestry manual
D Ratings based on soils memo 74, January 1972
* Site index is a summary of 5 or more measurements on this soil

Fig. 9. Part of an SCS Soil Survey Interpretation form. Shown (*from top to bottom*) are a description of a soil, its limitations for recreation, and its suitability for woodland and wildlife habitat. *Courtesy of the USDA Soil Conservation Service.*

cies on his or her land, especially when selling timber without a forester's assistance or evaluating the potential of a property for timber production. Several reference guides to tree identification are listed in appendix 2. It is important to own guides which use features in addition to leaves, especially buds, twigs, and fruit, for identification. Leaves are variable in form and those of seedlings and saplings are often unreliable for accurate species recognition. Furthermore, in New England the leaves of deciduous species are absent for half the year. For practical purposes, the twigs, buds, bark, fruit, and general shape of deciduous trees are the most reliable features for identification.

The forest products information given in table 3 lists the New England tree species most valuable for timber, as well as log specifications and standards. Differences in local markets and changes in supply and demand may significantly affect the value of a species and the extent of its use for a given product at any point in time. Product specifications vary among mills, and should always be established before harvesting. High-quality sawlogs of the more valuable hardwoods—oak, sugar maple, cherry, ash, yellow and white birch—are virtually always marketable. White pine, spruce, and fir sawtimber is usually marketable, but its sale value has historically been tied to the cyclical construction industry. Markets and values for the less valuable species vary locally and frequently.

In assessing timber value and describing tree size, the dimensions of interest are *diameter at breast height* (DBH), that is, tree width at four and a half feet above the ground, and the number of logs the tree can be expected to yield if harvested. For timber management, combined information about the age and height of the trees in a stand is of great significance because it is a measure of the quality of the site for growing trees. Diameter is not relevant to judging the growth potential of a site because trees grow slowly in diameter when they grow too close together; diameter is primarily a reflection of the crowdedness, or density, of a stand. *Unlike diameter, however, tree height is little affected by how close together the trees grow, and is most affected by the quality of the site—that is, by the supply of soil nutrients, moisture, and temperature.* A stand of tall, relatively young trees is an indication of an area where soil, temperature, and moisture conditions promote rapid tree growth, and where the highest economic returns will most likely be realized from investments in timber management.

The growth rate of a stand is also an indication of how long it will take

Fig. 10. Some signs of poor tree vigor or of decay, *clockwise from top left:* a seam, a canker, a large branch stub, a wound with conks, dead branches in the crown, and wilting leaves.

for the trees in the stand to reach a certain average size. Some ten-foot white pines in an overgrown pasture may or may not afford a middle-aged owner a commercial harvest within his or her lifetime, depending greatly on the quality of the site.

Trees having potential for high-quality timber are disease-free and straight with few, if any, defects such as wounds, seams, and large dead branch stubs. Disease signs to look for include conks (the fruiting bodies of fungi), cankers, dead branches in the leafy crown of a tree, leaves or needles that are not green in summer, and wilting leaves or exudations of sap where there is no wound; figure 10 illustrates some of the common defects and disease signs. A tree exhibiting these defects may not live long enough to make a sawlog. If it is already of sawlog size, its lumber is likely to be defective, perhaps even worthless.

Trees are valuable for veneer or sawtimber only if they contain a minimum of one nine-foot section along the middle of which an imaginary straight line will not run off into space (see fig. 11). A tree having only one such section is of value only if that section is in the lowest part of the tree, since the "butt" log contains the largest volume of high-quality wood. Trees with high sawtimber potential have several branchless, straight sections, and a full crown that constitutes from 20 to 30 percent of the tree's total length (fig. 12).

Table 3 Log Specifications and Relative Values of Forest Products

	Minimum Length of Log (feet and inches)	Minimum Diameter (inches)[a]	Standards	Species
		High-Value Products		
Veneer	8'6"	12"	Narrow limits on allowable defects and taper; requires straightness and no seams	Hardwoods: oak; black, paper, and yellow birch; basswood; ash; black cherry; sugar, red maple; beech; yellow poplar; aspen; butternut Softwoods: spruce; white pine; hemlock
Poles and pilings	10' to 50', according to product	5" to 11"	Narrow limits on allowable defects and taper; straightness critical	Hardwoods: white oak; black locust; hickory Softwoods: red pine; larch; white cedar, spruce
Hardwood sawlogs	8'4" to 16'4", in 1-foot increments	8" to 10"; often 12" to 14" for top grade	Number and location of defects strictly limited; log grades set by local standards[b]	Hardwoods: oaks; sugar, red maple; yellow, black, paper birch; black cherry; ash; beech; hickory; butternut; basswood; yellow poplar; elm
Softwood sawlogs	8'4" to 16'4", in 2-foot increments	6" to 8"; often 10" to 12" for top grade	Log grades set by local standards[b]	Softwoods: white, red, pitch pine; red, white, norway spruce; hemlock; balsam fir; larch; white cedar

Table 3—*continued*

	Low-Value Products			
Railroad tie logs	8'6" to 9'6", according to local standards	11" to 12"	Decay-free, fairly straight logs	Hardwoods: any except butternut and basswood Softwoods: none
Landscape tie logs	Not critical, but usually no less than 8'4"	8"	Rot acceptable	Hardwoods: any species; aspen preferred Softwoods: any except red pine
Pallet logs	Usually 8'4", but 4'4" at some mills	8"	Number and location of defects unimportant; fairly straight logs	Hardwoods: any species Softwoods: any species
Fuelwood	Cut to order	5"	Decay-free logs; straightness relatively unimportant	Hardwoods: most species acceptable; least desirable are aspen, cottonwood, basswood; most desirable are oaks, hickory, beech, sugar maple
Pulpwood	4', 5', 8', or tree length, according to buyer	5"	Decay-free logs; straightness relatively unimportant	Hardwoods: any species Softwoods: any species except cedar

[a] Measured at the small end of the log; excludes bark thickness.
[b] Although there are three sawlog grades for hardwoods and three for softwoods defined by the U.S. Forest Service, few New England mills use them. Most mills that grade logs use their own grading systems.

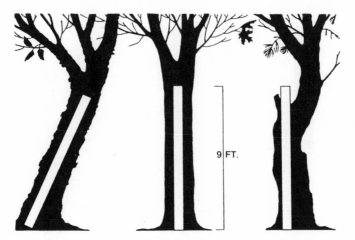

Fig. 11. Minimum standards for straightness in a sawtimber tree. The tree on the left and the tree in the center are straight enough to furnish a single sawlog, although the wood quality of the leaning tree would be inferior. The tree on the right is too crooked for sawtimber.

Fig. 12. Ideal crown size for timber trees. Crowns should be full and should extend from 20 to 30 percent of a tree's total length. Trees with crowns that are too small are usually slow growing; crowns that are too long result in knotty lumber.

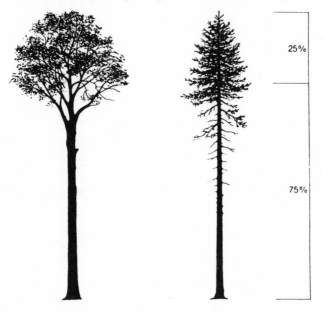

As a rule, New England forests have too many trees per acre for the rapid development of veneer trees or sawtimber. Each acre has a certain fixed productivity potential; if many trees share that capacity, growth in the diameter of each tree is small. Overcrowding diminishes diameter growth because it results in competition among trees, especially for light. However, overcrowding is not an indication of poor growth *potential*. When some trees are removed by thinning, their share of the growth potential is inherited by those that remain. If the trees removed during a thinning can be used as firewood or sold as pulpwood, overcrowding even becomes an asset.

Less common is woodland in which there are too few trees per acre. This situation may have implications for the land's productivity potential, and for the value of the trees that are currently there. Where trees are widely spaced there is little competition among neighboring individuals; lower branches receive a good deal of sunlight and persist longer, thus forming large knots in the butt log and reducing the quality of the timber. When considering a remedy for an understocked stand, which usually involves removing the existing stand and regenerating a new one, it is critical to know if the sparse stocking is a result of past land use or of an inherent inability of the land to support more trees. In the latter case it is, for all practical purposes, very difficult to improve the stand through management.

Maple Sap Production

Like timber management, sugarbush management is an ongoing process of establishing and cultivating selected trees. In sugarbush management, however, the goal is not maximum annual tree growth, but maximum annual syrup yield and operational efficiency.

In assessing the operability of a sugarbush, tree size is important: trees smaller than 10 inches in diameter will be permanently and seriously damaged by tapping. The size of a sugarbush is immaterial for landowners who sugar as a hobby. On the average, 1 tap will yield about 9 gallons of sap during a season, which will boil down to 1 quart of syrup. (Yields can be increased to around 15 gallons of sap per tap by using a vacuum-assisted collection system, as discussed below and in chapter 5.) A commercial sugaring operation requires substantial acreage. A well-managed sugarbush has between 70 and 90 taps per acre, which annually produce from 600 to 800 gallons of sap, or about 20 gallons of syrup. Syrup pro-

Fig. 13. Ideal crown size for a sugar tree. Because the leaves produce the sweetness in maple sap, crowns should be as wide and as long as possible. *Photo courtesy of the USDA Forest Service.*

duction is usually profitable only in operations having 1,500 taps or more and requiring, therefore, about 20 acres of good sugarbush.

Some landowners who do not do their own sugaring are able to sell sap or rent trees to syrup producers; under these arrangements no minimum number of taps or acres is necessary. The most critical factor in selling sap or leasing a sugarbush is finding a buyer or renter. It is easier to interest a producer in a small number of taps if they are at a roadside rather than in the woods. The price per gallon for sap, which is determined largely by its sugar content, is also affected by the amount of sap available to the buyer. It is most efficient for a buyer to transport large quantities of sap, usually 500 gallons or more, which usually bring higher prices per gallon. Since sap is highly perishable and difficult to store, it should be picked up by the buyer every two days; producing full truckloads of sap every two days probably requires a sugarbush of commercial size.

A potential commercial sugarbush should contain a sufficient number of promising sugar maple trees per acre. (Red, silver, Norway, and ash-leaved maple produce a sweet sap and can be tapped, but their sap is inferior to that of sugar maple.) The minimum number of sugar trees and the maximum number of taps per tree for a well-stocked stand sugarbush are:

Average DBH	4	6	8	10	12	14	16	18	20	22	24	26	28	30
Trees per Acre	167	126	98	79	64	54	46	39	34	30	26	23	21	19
Taps per Tree	0	0	0	1	1	1	1	1	2	2	2	3	3	3

See page 202 and Appendix 2 for further tapping guidance.

In sugarbush management, tree quality is related to potential sap production, and is therefore judged by different standards than those by which a tree is assessed for timber potential. A good sugar tree is most different from a timber tree in crown size, which should be maximized for sap production; compare the crown of the tree shown in figure 13 with the crowns of the timber trees shown in figure 12. Generally, the greater the leaf surface of a sugar maple, the more abundant and the sweeter the sap. Trees with crowns more than 30 feet wide can produce at least 100 percent more sap than trees with narrower crowns. In addition, the sap from large-crowned trees can be as much as 30 percent sweeter than that from trees with smaller crowns. Furthermore, large-crowned trees grow faster and therefore reach tappable size sooner than less vigorous trees.

A good sugar tree has a high sugar content. An increase of 1 percent in sap sugar content will reduce processing costs by 30 to 50 percent. Although absolute values of sap sugar content vary from hour to hour and by the day and the year, the relative sweetness of trees in a stand will remain constant; the sweetest trees in a sugarbush will always be so. The average sweetness of maple sap in New England is 2.5 percent sugar. Stands with an average sweetness of less than 2.2 percent are poor prospects for commercial tapping.

Because tree quality in a sugarbush is based on potential sap production crookedness and forking have no effect on the value of a tree unless the forks are acutely v-shaped and therefore structurally risky. Seams or wounds in the bole which have healed should not affect tree health and longevity. Cankers, large dead branches, rot, and other symptoms of disease or damage identify trees of declining vigor. Such a condition may in fact temporarily stimulate sap production, but will shorten the life of the tree.

A good growing site is very important to sugarbush management. On good sites, when older trees die, young trees replace them relatively quickly, and crown spread, which increases sap production and sweetness, can be maximized. Good sites encourage the rapid overgrowth of tapholes, thus minimizing tree decay.

The soils that are best for sugar maples are fertile, moist, well-drained loams, with a pH of 5.5 to 7.3. Heavy clay and sandy soils are unsuitable. A soil's pH is a measure of its acidity or alkalinity. The pH scale is from 0 to 14, with 0 being very acidic, 14 being highly alkaline, and 7 being neutral.

Terrain and accessibility can affect sap production and the efficiency of sap collection. Southern slopes lose their snow earliest and dry out more quickly, and there the sap runs first in the season and for a longer time each day. In most commercial sugarbushes and many hobby operations, sap is collected through plastic tubing attached to the trees and extended to a collection point (see fig. 14). Tubing systems work best on sloped sites. A 10 percent slope is ideal unless a vacuum pump is added to the tubing system. A pump replaces gravity as the force moving the sap through the tubing to the collection point and therefore reduces the need for a sloped site. Where buckets are used instead of tubing, steep slopes will cause significant delays in tapping and gathering.

Accessibility is as important for profitable sap production as it is for timber production. A dividing ridge that places sections of a sugarbush on two different slopes can make operation difficult; a valley may facilitate operations. Difficult terrain, or long distances between collection points and the sugarbush can reduce or cancel the value of the sap.

Fig. 14. Tubing for collecting maple sap. "Drop lines" connect the plastic taps in the tree to a tube line. The tube line collects sap from several trees, and feeds into a main line that goes to a gathering tank. Commercial sugar producers now use tubing systems instead of buckets.

Fuelwood

The sale value of fuelwood on the stump is generally from 5 to 20 percent that of sawtimber. From a management viewpoint, fuelwood should usually be treated as a by-product of other forest yields—that is, it should be produced from the low-quality trees removed to improve the wildlife habitat, timber quality, or sap productivity of a stand of trees. Except perhaps on tracts smaller than 10 acres, where sawtimber production may not be economical, management of a forest solely for fuelwood probably means that high-quality trees will be burned or will be sold at prices greatly below their potential value.

Because of the low unit value of fuelwood and the labor involved in processing it, difficulties of distance and terrain are especially significant when considering fuelwood production as a primary goal of forest management. Hauling fuelwood for more than 600 to 800 feet to get it out of the woods is uneconomical in most cases. Slope constraints are similar to those for timber. The common assessment is that the most accessible land is the best land for fuelwood production—the same land is normally the most expensive and that probably has potential for better returns from other uses.

Forest management, then, usually rates fuelwood production as secondary to timber, sugarbush, or wildlife habitat cultivation; trees relegated to the woodpile are those unqualified for timber or sap production, or with little value to wildlife. Where fuelwood is of interest not as a cash crop but as a source of heat for a landowner's residence, at least 10 acres of woodland, stocked primarily with hardwoods, are necessary. Typically, an unmanaged New England forest acre contains from 5 to 15 cords of potential fuelwood that have accumulated as the stand grew, and that should be removed to space the remaining trees for efficient wood production. (A cord is defined in appendix 1.) For a house that consumes 5 cords of wood annually, each acre initially represents roughly two winters' heat. However, if the land has been under management, or for some other reason is at the point where the initial supply of "excess" trees is exhausted, an acre can be expected to grow between a half cord and a cord of wood per year. This annual production can be harvested without decreasing the forest's productive base; for a sustained yield of fuel-wood, the same five-cord house will need a minimum of 5 to 10 acres of woodlot. Both the amount of excess wood and the average annual growth of trees in a lot can be determined by a forester.

In order to support the fuelwood needs of an average New England household, tracts smaller than 10 acres must be managed primarily and intensively for fuelwood, with less consideration given to timber, sugarbush, habitat, or aesthetic values.

For an owner wishing to make a commercial fuelwood sale, it is difficult to attract a buyer if less than 5 cords per acre are to be harvested. Very few owners of large equipment, such as skidders, will move their machinery to a job site for less than 100 to 200 cords. Smaller operators, using such equipment as farm tractors or pickup trucks, will sometimes purchase quantities as small as 40 or 50 cords.

Any tree species, including softwoods, makes acceptable fuelwood for airtight stoves or furnaces. The heating value of a species is rated by BTUs per cord of air-dried wood (see table 18). Table 4 illustrates how the fuelwood value of a particular species may be modified when characteristics other than heating value are considered. Elm, for example, has an average heating value, but it is difficult to split and therefore has a low overall fuel value.

If fuelwood is not a by-product, but is the primary management objective on woodlots smaller than 10 acres, cultivation of species that sprout readily from stumps or roots and that are relatively fast-growing is advisable. Because fast-growing species are typically less dense then slower-

Table 4 Fuelwood Value of Selected Woods

Species	Relative Heat Value per Cord*	Ease of Splitting	Ease of Starting	Sparks	Coaling Quality
Beech	High	Hard	Poor	Few	Excellent
Hickory	High	Medium	Fair	Moderate	Excellent
Hophornbeam	High	Hard	Poor	Few	Excellent
Oak	High	Medium	Poor	Few	Excellent
Sugar maple	High	Medium	Poor	Few	Excellent
Apple	Average	Hard	Poor	Few	Excellent
Ash	Average	Easy	Fair	Few	Good
Birch	Average	Medium	Good	Moderate	Good
Black cherry	Average	Medium	Good	Few	Excellent
Black walnut	Average	Medium	Fair	Few	Good
Elm	Average	Hard	Fair	Very few	Good
Larch	Average	Easy	Excellent	Moderate	Good
Red maple	Average	Medium	Fair	Few	Good
Aspen	Low	Easy	Good	Moderate	Poor
Basswood	Low	Easy	Poor	Few	Poor
Butternut	Low	Easy	Poor	Few	Good
Cottonwood	Low	Easy	Excellent	Moderate	Good
Fir	Low	Easy	Excellent	Many	Poor
Hemlock	Low	Easy	Excellent	Many	Poor
Pine	Low	Easy	Excellent	Many	Poor
Spruce	Low	Easy	Excellent	Many	Poor

Sources: C. Hunt and R. Ramath, *Enjoy Your Fireplace, Especially During the Energy Crisis* (Upper Darby, Pa., U.S.D.A. Forest Service, 1973); and assessments by the authors.
*Based upon BTUs per cord of air-dried wood.

growing trees a greater volume of their wood is needed to heat a home, but their high productivity means that heat output per acre per year is often greater than that from such dense slower-growing species as oak or beech. However, sprouts of even the slower-growing species will outgrow trees that originated from seed because the sprouts are supported by an existing, well-established root system. Except for pitch pine and perhaps northern white cedar, New England softwoods do not sprout. The region's hardwoods do, but with varying degrees of vigor. Aspen, ash, oak, basswood, red maple, and black cherry sprout vigorously. Beech forms numerous sprouts that do not grow very rapidly. Birch sprouts are often weak.

The density, or crowdedness, of a stand where fuelwood is being harvested affects the ease with which trees can be felled to the ground. Dense stands tend to have trees with fewer lower branches and smaller crowns;

these features make cutting, splitting, and stacking the logs easier. Lopping branches from trees with large crowns is both time-consuming and dangerous.

There is no strict maximum or minimum size for fuelwood trees. The optional diameter depends on what equipment is available for cutting, hauling, and processing. A landowner lifting 4-foot logs onto a trailer or pickup truck would probably prefer trees from 5 to 8 inches DBH, but heavy skidding equipment cannot deal efficiently with trees of less than 8 inches in diameter. The following table indicates the approximate number of trees of selected diameters needed to make a cord of wood on an average New England site:

DBH (inches)	5	6	7	8	9	10	11	12	14	16	22
Approximate Number of Trees per Cord	46	21	15	10	8	6	5	4	3	2	1

Clearly, the cordage of larger trees mounts up very quickly. However, large trees are more dangerous to fell, and more of their wood will require splitting.

Tree quality is unimportant in fuelwood, although rotten wood does not make acceptable firewood. Large knots and other defects are significant only to the extent that they make splitting more difficult.

Pulpwood

Pulp for papermaking has been traditionally made from trees between 5 and 12 inches in diameter, cut into four-, five-, or eight-foot lengths called *sticks*. Pulpwood is sold by the cord, cubic foot, or ton, either directly to paper or pulp mill or to a logger who sells it to a mill. At the mill, the logs are reduced to small chips as the first step in making wood pulp. In recent years, delivery of whole trees (not logs) to the mill and the chipping of trees right in the woods have become increasingly common practices, especially on land owned by forest industries.

Most of the wood pulp mills in New England are located in northern New Hampshire and Maine. Mills in Quebec purchase pulpwood from northern New England, and a few mills in eastern New York State import pulp from western New England. Outlets for pulpwood in central and southern New England are beginning to develop, although the end product is not usually paper.

Much of the pulpwood harvested in New England comes from land owned by paper companies. On nonindustrial private land the stumpage value of pulpwood is similar to that of fuelwood and, like fuelwood, it is usually best considered as a by-product of management for other uses. Profitable pulpwood production requires a nearby market and good access to the stand.

Because pulpwood is a low-value product, many of the constraints that apply to fuelwood production also apply to pulpwood. Most pulpwood operations involve large equipment, so a significant volume of wood per acre and a large total volume must be available; 5 cords per acre and an overall volume of 100 to 200 are bare minimums. Operators who use highly automated machines and chippers (see chap. 6) probably require substantially larger volumes. If chipping is done on the site, the landing area has to be large and relatively level to accommodate the chipper, log piles, and chip vans associated with a mechanized operation.

Softwood species are most desirable for pulpwood; however, an increasing proportion of New England's annual pulpwood production is in hardwoods, which are in more plentiful supply. Spruce and fir are preferred species for papermaking. Tree form is of little concern in pulpwood quality, although very crooked sticks are unsuitable because they cannot be processed in the chipping machinery. Pulpwood should be free of rot.

Note that whereas "pulpwood" sometimes is used also to refer to industrial fuelwood, its use here is restricted to the raw material for papermaking.

Wildlife Habitat

On private land in New England, wildlife management means creating or improving wildlife *habitat*—that is, the combination of food, shelter (cover) and water that will attract and support animals. Wildlife management does not mean feeding animals, releasing them, or eliminating their predators (see chap. 5). These practices usually work to the detriment of the animals they intend to help.

There are two ways to approach habitat management. In the first, the goal is to maximize the diversity, or variability, of forest vegetation by ensuring the presence of several heights, ages, and species of trees, and both forested and open areas. Most wildlife is best able to use diversity when forested and open areas are interspersed rather than in large, uniform blocks. Interspersion creates the largest possible amount of *edge*, the transition area where field and forest, or different forest stands, meet and grade into one another (see fig. 15). Transition areas usually provide dense cover and a variety of different plants. Ideally, streams, ponds, or wetlands will also be available. An increase in vegetational diversity in a forest permits a wide range of food and shelter, and expands the number of animal species that will be attracted.

The second approach to habitat management is directed toward fostering a single species. Efforts are made to increase the resident population of that species by carefully creating its ideal habitat; other species are only of secondary consideration.

The maximum-diversity approach to habitat management requires no minimum acreage. It aims to attract the widest possible variety of animals to a piece of land. Much can be done at minimal expense to increase diversity on tracts of any size. Success with the species-specific approach, however, may require large acreage, depending on the species. An individual animal of a given species has a *base range* within which food, cover, and water must be present in adequate quantity and quality; desirable habitat for one species may be undesirable for another. The typical base ranges for some native species of New England wildlife are given in table 5. To the extent that the necessary habitat elements are not available within a base range, an animal must go farther afield than its minimum range. If a landowner's tract provides less than the minimum base range for a species of interest, the owner can evaluate adjoining properties to see if all required habitat elements can be supplied through cooperative management.

Soils are important in habitat management to the extent that they permit or limit vegetational diversity. Droughty soils especially are poorly suited to supporting large quantities of herbaceous and shrubby vegetation. It is a mistake to relegate wildlife management to the poorest soils and sites. The most productive land for timber is also likely to be the best land for supporting wildlife, and often the two objectives can be achieved on the same site.

Increasingly threatened by development, wetlands are essential ecosystems and critical wildlife habitat. *Every effort should be made to avoid their disturbance and they should be surrounded by protective buffer zones* (See appendixes 3 and 4). Although ponds can attract many of the same types of wildlife as do wetlands, the construction or impoundment of

Fig. 15. Edge. The margins between different types of vegetation, or between land and water, are called edge and are rich wildlife habitats. In the photo are at least two edge areas: the brushy field-forest border and the boundary between the softwood stand and the hardwood stand.

Table 5 Estimated Size of Base Ranges of Selected Wildlife Species

	Size of Range in Good Habitat
Cottontail rabbit	½ acre
Squirrel	1 to 2 acres
Snowshoe hare	10 acres
Ruffed grouse	40 acres
Woodcock	50 acres
White-tailed deer	640 acres
Wild turkey	1,000 acres
Bobcat	6,000 acres
Moose	5,000 acres
Black bear	Unknown; estimated at 10,000 acres

Source: R. M. DeGraaf and D. D. Rudis. 1986. *New England Wildlife: Habitat, Natural History, and Distribution.* U.S.D.A. Forest Service GTR NE-108.

a pond on a wetland site will eliminate most of the wetland's other ecological values.

Assuming its construction will not affect a wetland, a pond will greatly increase the potential of a property to attract wildlife. The ability of a particular location to support a pond is a function of topography and the underlying soil. The SCS will assist a forest landowner in evaluating the feasibility of a pond in a non-wetland area and will provide engineering details.

In some instances, the slope and aspect of the land are important in wildlife habitat management. The preferred wintering areas of deer are softwoods, such as white cedar and hemlock, at middle and lower elevations on southern slopes. These areas are warmer than other terrain, offer greater protection from winds, and accumulate less snow. Aspen, a tree which provides excellent food and cover for a number of wildlife species, regenerates best on sunny, south-facing slopes. Steep slopes may restrict wildlife management if equipment used in habitat improvement is unable to operate on them.

Accessibility to a property may affect the potential for wildlife management in several ways. Difficult or seasonal access restricts an owner's ability to enjoy observing resident wildlife populations. On the other hand very timid species, such as wild turkey, are more likely to be present in areas where access is poor and human presence is minimal. Poor access also increases the costs of using equipment. Improving access to a property by the construction of roads can benefit wildlife as well as the owner. Seeding roads to a mix of grasses, legumes, and other herbaceous plants adds a food source used by insects, small mammals, and birds, including ruffed grouse and turkey.

The values of some New England tree species as wildlife food are listed in table 6. These food values are general, and will vary with the time of year and the combination of foods available at a site. It is useful to compare this list with forest product values given in table 3 if management for both timber and wildlife is being considered.

Trembling and bigtooth aspen are valuable for both food and cover. These closely related and extremely fast-growing species can be induced to regenerate in very high density, providing browse and thick cover for deer, grouse, turkey, woodcock, hare, and rabbit. Because aspens do not reproduce from seed nearly as readily or as profusely as they sprout from roots after clearcutting, it is critical to have vigorous, large aspens within an area to be regenerated. Clearcutting and other regeneration techniques are discussed in chapter 5.

Other important sources of food for wildlife are *mast* trees—those that produce fruit, nuts, or seeds eaten by a variety of woodland wildlife species. Perhaps the most important of the mast producers are the oaks, beech, hickory, ash, apples and hophornbeam.

Softwood species that are important for cover (especially for deer) include hemlock, cedar, spruce, fir, and pine. To provide adequate cover softwoods should constitute at least 50 percent of the overstory trees; stands should be at least 5 acres in size. To assess the importance and potential of an area as a deer yard, walk through the stand when the ground is snow-covered looking for tracks, snow beds, droppings, and evidence of browsing. *Deer yards are critical elements of deer habitat that should be managed with utmost care.*

In the maximum-diversity approach to habitat management, it is essential to aim at a variety of tree sizes, as well as species. Many woodland properties in New England are dominated by stands of the same size and age. This can be corrected by patch cuts and firewood thinnings. After cutting, hardwood species will sprout and herbaceous vegetation will occupy the ground level, providing food for browsing animals such as deer. For cover, tree species may not be as important to wildlife as tree size and shape. Particular wildlife species have particular preferences for tree form, but a mixture of dead and live, thick-crowned and feathery, tall and low trees will attract a diversity of species.

Table 6 Relative Values of Various Tree Species as Wildlife Food

Species/Food Value	Part of Tree Eaten					
	Buds	Twigs	Bark	Foliage	Mast	Catkins
Very High						
Oaks	X	X	X	X	X	
Black cherry	X	X		X	X	
Apple	X	X	X		X	
White pine	X	X		X	X	
Dogwood	X	X		X	X	
High						
Maples	X	X	X		X	
Aspens	X	X	X	X	X	
Serviceberry	X	X	X	X	X	
Pitch pine	X				X	
Hemlock	X	X	X	X	X	
Mountain ash	X	X		X	X	
Beech	X	X		X	X	
Medium						
Birches	X	X	X	X	X	X
Spruce			X	X	X	
Hickories		X	X	X	X	X
Balsam fir		X	X	X	X	
Alder	X	X		X	X	X
Tupelo	X	X		X	X	
Mulberry					X	
Elms	X	X		X	X	
White cedar		X		X	X	
Red cedar		X		X	X	
Willow	X	X	X	X	X	X
Yellow poplar		X		X	X	
Holly		X		X	X	
Hawthorne	X	X		X	X	
Black walnut		X			X	
Butternut					X	
Ashes		X		X	X	
Cottonwood	X	X	X	X		X
Mountain laurel	X	X		X	X	
Red pine	X				X	
Staghorn sumac		X	X	X	X	
Yew				X	X	
Low						
Hackberry		X		X	X	
Hophornbeam	X	X		X	X	X
Larch	X	X	X	X	X	

Table 6—*continued*

Musclewood	X	X	X	X	X	X
Rhododendron	X	X		X		
Sassafras	X	X	X	X	X	X
Norway spruce			X	X	X	
Red spruce			X	X	X	
Sycamore		X			X	
Witch hazel	X	X	X	X	X	X

Source: R. J. Gutierrez et al., *Managing Small Woodlands for Wildlife,* Information Bulletin no. 157 (Ithaca, N.Y.: Cornell University, 1979), p. 23.
Note: These food values are general, and will vary with the time of year and the combination of available foods at a site. All species listed are used by both mammals and birds, except for black walnut and butternut, which are used exclusively by mammals.

A critical aspect of an ecologically balanced forest is the presence of tree cavities and rotten trees, which provide food, nesting, and denning opportunities for a variety of birds, mammals, and reptiles. Dead or dying standing trees, called *snags*, are valuable wildlife habitat elements; the best snags are those of more than eight inches DBH that still have bark (fig. 16). The retention of snags has only a minimal effect on the efficiency of timber production.

High tree density can have both positive and negative impacts on an area's potential for supporting wildlife. Where cover is the primary concern, as in winter deer yards, dense stands of trees provide greater shelter from wind and significantly reduce snow depth on the ground. Dense young stands of some species, such as alder and aspen, provide nesting opportunities and a degree of protection from some predators. Where food is the main concern, less crowded stands are generally more productive insofar as crowns of mast trees have a chance to expand with minimal competition from neighboring trees.

Stands with fewer trees per acre in the overstory encourage the development of several vertical layers of vegetation, since more sunlight reaches the ground.

In sum, the potential of a property for wildlife management is not fixed. The manipulation of vegetation to increase variety of species, ages, and sizes is practical on almost any property where the maximum-diversity approach is selected. The species-specific approach is tricky, and a landowner should probably seek professional assistance before beginning an involved

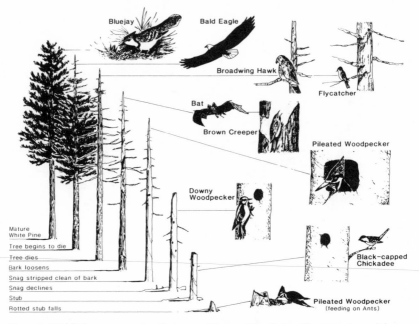

Fig. 16. Wildlife use snags in all stages of decay. The figure illustrates how different species of birds, for instance, will use a tree as it deteriorates. Often the wildlife function of a snag is simply as a perch that provides visibility. Availability of such perches alone may largely determine whether a bird species is present or absent in a forest. *Source:* R. M. De Graaf and A. L. Shigo, *Managing Cavity Trees for Wildlife in the Northeast,* U.S.D.A. Forest Service General Technical Report NE-101 (Washington, D.C. U.S. Department of Agriculture, 1985).

program aimed at one species. In both approaches, it is important to realize that property boundaries are meaningless to wildlife. The ecosystem that supports or discriminates against wildlife is much larger than any single parcel of woodland.

Fragile Areas

Swamps, marshes, bogs, riparian edges, and very soggy meadows and woods, all known as wetlands, are rich ecosystems and wildlife habitat. Until recent times, wetlands were considered unproductive wastelands. They have been flooded, drained and filled extensively resulting in significant ecological damage and species loss. Proper protection of a wetland will require limits on an owner's activities in its vicinity and some protection of significant wetlands is required by federal and state laws (see appendix 3).

However, the prospective buyer interested in the wildlife and aesthetic values of a piece of woodland should consider the presence of a wetland as an asset offering unmatched opportunities for wildlife and ecological observation. Wetlands should not be regarded as potential pond sites.

Similarly, the presence of rare or endangered plant and wildlife species should be considered as both an added responsibility and an extra asset. Plants and animals whose existence is threatened globally or nationally are listed and protected by the U.S. Fish and Wildlife Service. These species have the strongest levels of protection. Species considered rare or endangered only in the New England region or one of its states are listed and protected to varying degrees by state laws. Some New England states do not require the protection of state-listed species during forestry or agricultural activities (see appendix 3).

Recreation and Aesthetics

Many New England landowners do not hold their land primarily for production or for the income it might generate. Surveys have shown that the less tangible benefits of privacy, recreation, and aesthetic enjoyment are of most importance to many of them. These values are largely personal, and to that extent are difficult to discuss, but a little guidance can be offered.

Hilly but gentle terrain is preferable for most forms of forest recreation, and especially for hiking and cross-country skiing. In addition, varied terrain makes for a diversity of vegetational communities that is aesthetically pleasing. Changes in altitude offer opportunities for long-range views. A topographic map may reveal areas where trees can be cut to give a desirable vista.

Streams are valuable aesthetic features to many landowners. Since many New England streams are intermittent—full of water only in the spring and briefly during other wet periods—it is important to be sure that a stream runs year-round before considering it as a major woodland asset.

Species diversity within a forest is often aesthetically pleasing to a forest landowner, although this is a very personal value. White birch and beech are well known for their beauty. Maples are key members of the fall foliage display. Mixed softwood and hardwood stands are especially pretty in the winter when the needles of the softwoods add a welcome green to the dormant forest. Most people prefer large trees to small; an owner must evaluate the timber and wildlife benefits given up when trees are retained solely because they are large.

Very dense softwood stands, especially spruce and fir, are difficult to walk through and see into. Many people like the look of open, parklike stands. Such an effect can be created by thinning, but it is important to realize that sprouts and ground vegetation will soon occupy an area after a thinning, and that the stand may become denser than it was originally. Repeated clearing of the understory may be necessary to maintain an open appearance, but will negatively affect wildlife.

Simple Techniques for Assessing Potential

A landowner can informally and inexpensively estimate or evaluate many of the features of a parcel of woodland with the methods and tools described in this section.

If the boundaries of a property are described by their length and compass bearings in a deed or on a survey map, the size of the tract can be easily calculated by tracing the map exactly onto graph paper, as shown in figure 17. By taking into consideration the scale of the map and counting the number of squares falling within the drawn property boundary, a fairly accurate estimate of the area can be made. Figure 17 presents such a diagram for a 155-acre property. Assume that on the grid of the graph paper, 5 squares in a row are 1 inch long and that, on the map, 1 inch equals 500 feet on the ground. When the map is traced onto the graph paper, one side of a square is equal to 100 feet, and the area of the square is 10,000 square feet. Estimating acreage involves:

1. Counting all of the whole squares included with the boundary, and every other square that is only partly included; in figure 17, 667 squares are counted
2. Multiplying the number of squares by the area represented within one square; in figure 17 that area is 10,000 square feet; $667 \times 10,000 = 6,670,000$ square feet
3. Converting the product of multiplication to acreage; in the example, this would be done by dividing by 43,560, the number of square feet in an acre, which yields a little more than 153 acres—a close approximation to the acreage determined by a professional survey.

Tract size can also be calculated using aerial photographs or topographic maps, although the assistance of a forester will probably be necessary for the interpretation of aerial photos. Some pertinent references are listed in appendix 2.

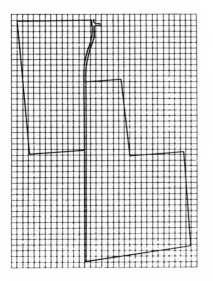

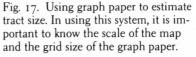

Fig. 17. Using graph paper to estimate tract size. In using this system, it is important to know the scale of the map and the grid size of the graph paper.

A *compass* is useful in forest management for determining aspect, for locating boundaries, and for navigating in the woods (fig. 18). A compass needle always points north—not to the North Pole, but to a magnetic mass of rock in northern Canada. The direction to the North Pole is called "true" north, and the direction in which a compass points is called "magnetic" north. The number of degrees that the compass needle deviates is known as "declination." The amount of declination varies from place to place, and changes over long periods of time. When looking at a map or a deed description of a property, it is important to know three things: (1) the declination at the location of the property (declination ranges from about 14 degrees to about 21 degrees west in New England; (2) whether the bearings given on a map or in a deed are true or magnetic; and (3) whether the declination has changed appreciably since the bearings were determined. Changes in declination since surveying began in New England have been minor. The declination at any given location within New England can be obtained from the bottom of a U.S. Geological Survey topographical map on which the property is included (see fig. 19). Most boundary maps and deeds, and especially older ones, use magnetic bearings.

Landowners who have been lost on their own property can attest to the wisdom of owning a compass, regardless of its usefulness for mapping and boundary work. There are many compasses of fine quality on the market. A liquid-filled housing is desirable since it allows the needle to settle

Fig. 18. A liquid-filled compass with a sighting mirror, declination correction arrow, and a clinometer.

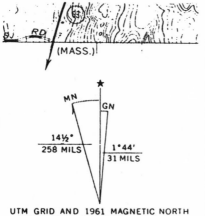

UTM GRID AND 1961 MAGNETIC NORTH
DECLINATION AT CENTER OF SHEET

Fig. 19. Declination information as it appears on a USGS topographical map. This information appears left of center on the bottom margin of all USGS topographical maps. Of interest to most map readers is the line topped by the star, which indicates true north on the map, and the arrow marked "MN," for magnetic north. To the left of the magnetic north arrow the declination for the area included in the map is given in degrees (here, 14.5°) and mils. Declination is commonly expressed in degrees.

quickly. The degree readings on a compass should be graduated finely enough to be useful for the fairly precise work of boundary location; graduations of two degrees are sufficient. Some compasses automatically correct for declination, saving the user from having to calculate each reading where true bearings are necessary. It makes sense to purchase a compass that can be used for several tasks. A compass mounted on a clear plastic base that is marked off in inches or centimeters is useful for measuring distances on a map when in the field. Clinometers, which are simple and inexpensive tools for measuring steepness, are built into some compasses (see fig. 18).

Distances can be estimated by pacing. To calibrate one's pace for walking in the woods, a 100-foot distance is measured on level ground with a measuring tape; that distance is walked at a normal stride several times, while the number of paces on each walk is counted. By convention, one pace is counted each time the same foot is used; two steps equal one pace. The average number of paces for the 100-foot walk is then calculated, and that number is divided into 100 feet to determine the average length of one's pace. A distance in the field can then be estimated by walking over it while counting paces, then multiplying the number of paces by the length of one's pace.

It is important to understand that the distances referred to on maps is *horizontal distance*, which is less than *sloped distance*; these concepts can be seen in figure 7. It is clear that the horizontal line along the bottom of the figure is shorter than the lines representing 10 and 40 percent slopes. A 100-foot measurement on a map or a deed description means 100 feet of horizontal distance, and therefore the actual distance to be traveled is probably greater than 100 feet. When estimating distances in the woods by pacing, adjustments must be made for walking up or down a slope,and for the fact that pace length tends to vary going uphill and downhill. References on mapping and surveying in appendix 2 describe ways of making these adjustments.

A *tree scale stick* is the least expensive device for estimating tree diameter and height. Tree scale sticks resemble yard sticks, and are available from some local hardware stores and from forestry equipment suppliers listed in appendix 4. Directions for their use are printed on the stick.

Without cutting a tree down and counting the growth rings on the stump, there is no means of accurately determining tree age unless an expensive and sometimes damaging piece of equipment called an *increment borer* is used (figs. 20, 21). However, it is possible to develop an eye for external tree features that indicate youth or maturity. The most useful of these are bark characteristics, since virtually all New England species have a smooth bark in youth which becomes increasingly ridged, flaky, or furrowed with age. A small tree with "old" bark is probably a slow-grower, and a stand of such trees may indicate a poor site for growing timber.

Figs. 20, 21. Using an increment borer. The threaded borer tube, which is hollow, is drilled into the tree (fig. 20), and the core of wood is withdrawn from the tube (fig. 21). The core shows the tree's annual growth rings and provides an estimate of its age. Although indispensable in forest management because it is the only tool for determining tree age, the increment borer should be used sparingly because it can cause significant damage. *Photos courtesy of Forestry Suppliers, Inc.*

Measuring sap sweetness with a *refractometer* is a straightforward process: a drop of sap is placed on the prism of the instrument, and the sugar content can be read directly on the scale that is seen through the eyepiece (fig. 22). The process of choosing trees from which to sample sap is described in several of the references on sugarbush management listed in appendix 2. Refractometers are expensive to purchase. County foresters may have information on the availability of refractometers that can be borrowed.

Some of the information that is important in assessing woodland potential is available in publications, such as the soils maps and descriptions published by the Soil Conservation Service. In many New England states, the Extension Service or the forestry agency of the state government publishes reports of current prices being paid for trees, which help landowners to become familiar with timber values (see appendix 4). Assessments of a few of the forest features discussed in this chapter require the assistance of a forester. Estimating the degree of tree crowdedness, for instance, requires skill with specialized forestry tools and an experienced eye.

Compatibility of Uses

The choice of forest uses on the basis of woodland potential can be further refined by considerations of compatibility. Where multiple use of the same site may be especially important, as it is on woodlots smaller than 25 acres, the effect of management for one purpose on other values must be assessed. An acre of woodland may have excellent potential for two uses

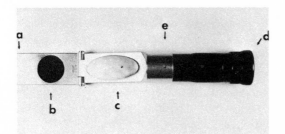

Fig. 22. A refractometer. A drop of sap is placed on the plate (b) which is on the cover (a). The cover is closed so that the sap is spread on the prism (c). A scale indicating the sap sugar content as a percentage is visible through the eyepiece (d). The instrument is calibrated by means of a screw (e).

that conflict. In practice, however, only a few uses are incompatible, and most can be made to coexist through compromises (see the summary of compatibilities in fig. 23).

Incompatible uses can be defined as those which require such extensive compromises with one another that the returns frome each are minimal. Two of the least compatible combinations are discussed below: sugarbush management and wildlife habitat improvement, and timber production and sugarbush management.

It is difficult to maintain wildlife habitat in a sugarbush for two reasons: first, there is no tree species diversity; second, sugarbush management requires the eradication of understory vegetation to facilitate sugaring operations and to reduce root competition. In larger sugarbushes, species limitations might be mitigated to some extent by the retention of scattered individual trees of other species with only minor losses in sap production. It is not as easy to compromise the need to eliminate the forest understory, since an understory is an impediment to sap collection, keeps microclimatic temperatures low in the spring, and may reduce sap production by allowing root competition. Removal of the understory, however, takes away forest cover and food sources vital to many forms of wildlife.

There are several ways in which sugarbush and timber management conflict. Because the best sap production and sugar content are usually obtained from trees with wide, deep crowns, intensive sugarbush management necessitates a wide spacing of trees that encourages branching; such branching is often excessive for the production of good lumber. Timber production becomes secondary to sugar production when a tree continues to be a good sap producer long after it has passed the age of maturity as a timber tree, and even after it has begun to decay. The portion of a sugar tree most likely to be "clear" (branchless) is the lowest (butt) log, but the lumber from this portion may have been damaged by repeated tapping, especially if antibiotics were applied in the tapholes. Antibiotics retard the process by which the tree seals off taphole wounds, and thereby work to extend the sap run. Their use is discouraged because they promote extensive wood decay around tapholes. Loggers and sawmills are reluctant to purchase logs from sugarbushes where metal spouts have been used since a forgotten spout imbedded in the wood is a threat to safety and to saw blades. The lower log of trees sold from former sugarbushes should be scanned with a portable metal detector before being harvested and sawn.

Recreational use of the land can be temporarily incompatible with hab-

Fig. 23. Common uses of forest land in New England and their compatibilities on the same acres. Compatible uses have a rating of 1; 2 indicates uses that are compatible if some compromise is made in each; combinations of uses that may be temporarily incompatible are those rated 3; and uses with a rating of 4 are not compatible on the same acres.

	TIMBER PRODUCTION	WILDLIFE HABITAT IMPROVEMENT	SUGARBUSH MANAGEMENT	FUELWOOD AND PULPWOOD PRODUCTION	RECREATION
WILDLIFE HABITAT IMPROVEMENT	2				
SUGARBUSH MANAGEMENT	4	4			
FUELWOOD AND PULPWOOD PRODUCTION	1	2	1		
RECREATION	3	3	1	1	
AESTHETICS	3	3	1	1	1

itat and timber management. Hiking, skiing, and even hunting are inhibited for up to ten years by habitat management which requires small clearcuts to regenerate shrubby cover for such species as ruffed grouse and woodcock. In clearcut areas there is likely to be *slash,* or the unused portions of trees, which makes walking difficult or impossible except on logging roads. The logging of sawtimber may similarly restrict the recreational use of land for five to ten years. Softwood slash persists on unshaded sites for many years unless it is cut into short lengths to hasten rotting. Hardwood slash is quicker to decompose, but where hardwood has been clearcut the density of regrowth may continue to make recreational use of the site difficult even after the slash has decayed (see the young aspen in fig. 57). Piled slash is extremely slow to decompose, and piling is labor intensive, but birds and many small mammals use such brush piles for cover. Slash disappears most quickly where it is in shade and where it has been lopped, or cut into short lengths so that it lies close to the ground; under these conditions, it may cease to be an impediment in as little as two years (see figs. 24, 25).

Some uses of forest land are highly compatible; not only do they have a small effect on each other, but in some cases the combination of uses improves the economic efficiency of overall woodland management. Fuelwood and pulpwood, for examples, are useful by-products of both sugarbush and timber management. Land used for recreational purposes can also be

Figs. 24, 25. Untreated and lopped slash. Slash, the tops and large branches of trees left after logging, presents a visual problem that can be minimized if the debris is cut into short sections (lopped) so that it lies close to the ground. *Photos courtesy of the Extension Service.*

used as a source of fuelwood or pulpwood with little impact on recreational potential because the trees cut for these products are usually scattered and small, they are utilized fully enough to eliminate much of the debris and they can be removed from the woods with small equipment, which often has only minimal effects on the appearance of the forest.

Sugarbushes are usually inviting for recreation because of the open, parklike spacing of the trees and the network of roads necessary for sugaring operations. Motorized recreational vehicles (except perhaps snowmobiles) should not be taken into sugarbushes, since sugar maple is extremely sensitive to soil compaction.

Some uses can be combined if careful compromises are made which only slightly reduce the benefits of each. The best examples are sawtimber production which improves wildlife habitat and, conversely, wildlife habitat improvement which is adapted to the economic considerations of timber value. In the first case, timber production is modified to provide food or dens for wildlife by allowing some of the growing space to be used by trees of unmerchantable species or form, by encouraging the growth of an understory (which might slightly deminish tree growth), or by foregoing the harvest of some economically valuable trees that provide important wildlife habitat elements. Patterns of harvesting can be designed to maximize the transition zones between different successional stages; such edges are the most productive areas for wildlife.

Where habitat management is the primary, but not the only, objective, the intensity of habitat management is determined by balancing the value of the available timber against the cost of habitat improvement. For example, while small clearcuts to improve wildlife habitat might ideally be scattered throughout a forest tract, they can be limited instead to areas where marketable timber will pay for the cutting. In such a scheme neither habitat nor timber values are maximized, but both are realized to some extent.

Management for aesthetics may be incompatible with timber production because bushy crowns and multiple-stemmed trees are beautiful, but they are shunned by timber buyers. A compromise can be made by growing a few beautiful, unmerchantable trees near frequently visited spots.

Comparative Economic Returns

It is impossible to quantify the economic benefits of various forest uses because they vary constantly, particularly in regard to different tracts and regions. In forest management, timber production is generally the most financially profitable use of an acre of woodland with reasonable potential. The economic outlook for the production of high-quality timber is very bright. A diminishing supply of high value species, especially red oak and ash, has led to an impressive rate of value growth during the last quarter of this century. As figure 26 shows, an increment of diameter growth in a large tree contains much more wood than equal growth in a smaller tree. Each increment in the length of the branchless stem in a good tree also increases the amount of high-quality wood. To date, increases in the value of poor-quality timber from unmanaged stands have often not kept pace with inflation.

The exception to the economic advantage of timber production over other forest uses may be sugarbush production, which is likely to yield more net income per acre than timber. *This holds true only if sugarbush management began early in the life of the stand, and if the stand is of commercial size.* Unlike sawtimber management, sugarbush management requires an annual labor input. On the other hand, after an initial period of establishment, about 25 to 30 years, a sugarbush yields an annual income, whereas a timber stand does not. The annual profits from sugaring are taxed as ordinary income. A stand of maple timber begins to yield sawlogs from thinnings when it is about 40 to 50 years old; income from this operation, which is usually eligible for capital gains treatment, might be anticipated every 10 years after that until final harvest.

Unless they are coupled with timber sales, ventures for wildlife, aesthetic, and noncommercial recreational purposes are clearly deficit operations. The costs associated with these undertakings are the expenses of hiring a logger or excavator to implement plans for improving habitat and views, and to construct roads and trails.

Some Advice on Buying Woodland

Prospective buyers of New England woodland can use the preceding sections of this chapter as a guide to evaluating and comparing tracts of land. There are several other important points that should be kept in mind during the process of finding and purchasing forest land.

New England woodland is a very good investment for the buyer who is looking for a hedge against inflation, a place for recreation and privacy, a basis for self-sufficiency, or a legacy for his or her heirs. Because forest

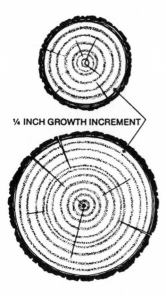

¼ INCH GROWTH INCREMENT

Fig. 26. The same annual diameter growth adds much more wood (and value) to a larger tree.

management can greatly augment these values, and can offset the costs of owning land, it is important to consider management potential when looking at woodland. All else being equal, land stocked with good timber will be worth much more than poorly managed land, especially since the value of good timber has historically appreciated faster than inflation. If wildlife habitat, sugarbush quality, and the aesthetics of a piece of land have also been improved by forest management, its value has been enhanced even further.

Forest land and forest management are usually not good investments if a buyer's sole intent is the generation of profits from timber, fuelwood, or maple products. The cost of land is generally based on its high value for commercial or residential development, and not on its much lower capacity to produce income from forest products. It is therefore virtually impossible to recoup land costs based on fair market values through sound forest management alone.

Sound forestry excludes speculative land purchases for the purpose of liquidating merchantable timber and then reselling the land. Such practices are often economically rewarding in the short run, but are a throwback to our history of using the forest without acknowledging its limits. In the long run, the practice of "cut-out-and-get-out" forestry has been and would again be economically disastrous for New England.

The high cost of land is especially discouraging if woodland is to be purchased solely as a source of fuelwood. Processing and selling fuelwood is a very difficult business which offers only marginal profits. Although the selling price of a cord of fuelwood is ten times or more than the wood is worth on the stump, the actual profit margin is very small because of the considerable labor, handling, and transportation involved in fuelwood operations. Selling log-length fuelwood entails the use of heavy equipment, the costs of which can cancel the savings in labor and handling. Selling fuelwood on the stump to a logger does not usually generate significant income, and allowing homeowners to cut their own wood for a fee presents potential liability risks for the landowner. In sum, expectations of making a profit from the purchase of land for fuelwood are unrealistic. If woodland is purchased as a source of fuel because it contributes to an owner's self-sufficiency, or is considered a hedge against future fuel cost increases, standards of current economy become much less important.

It is worth repeating that these discouraging words apply only to the economic prospects of purchasing New England woodland at fair market value, and solely for financial return from forest products.

When a buyer is seriously considering a piece of woodland a forester can be hired to provide a preliminary evaluation of the property's potential for the uses in mind. Such an assessment should take between a half-day and a day of the forester's time; if a detailed inventory is called for a few additional days will be needed.

Although real estate brokers are experts in many land values, few are familiar with forestry concerns in assessing woodland. If any forestry information is included in a real estate listing, it is usually a forester's appraisal of the financial value of the merchantable timber on a tract. It is important to interpret an appraisal correctly, and to distinguish between total value of the timber and the value of the timber that is currently reasonable to harvest. *Rarely, if ever, does the removal of all and of only the merchantable or mature timber from a piece of land constitute good forestry.*

Often land boundaries are not clearly discernible, and it is not unusual for a buyer to discover that the boundaries have been in dispute for several years. Vague deed descriptions are common, and actual acreage may not correspond with either the deed or the local assessors' records. One of the most important steps in purchasing woodland is obtaining a map of the property, and making certain of the location of the corners and boundaries (see the mapping and surveying section in chap. 4). Maps and related documents should be requested from the previous owner and checked for accuracy. It is dangerous to trust old deeds or maps that were prepared without the benefit of modern surveying techniques. Walking around the boundaries of a property under consideration is always wise. The advice of an attorney or a forester should be sought concerning the need for a legal survey by a registered surveyor, which can be a significant expense.

In a title search, a buyer's attorney should:

1. Establish that there is legally assured access if the property is landlocked;

2. Discover any rights-of-way through the property that are held by others;

3. Check for any liens and unpaid taxes on the land and the timber, and determine if the use of the land or the management of the forest is restrained by deed restrictions or so-called "use-value" or "current-use" state taxation programs;

4. Inform the buyer if any rights, including timber rights, are not being deeded with the land; all rights to use and develop a piece of land, including those to timber and minerals, can be bought and sold separately so it is possible for a previous owner to have sold or to retain timber rights. A conservation easement or conservation restriction on a piece of land means that some rights to develop it have been removed in order to protect its ecological, aesthetic or natural resource values.

If possible, a buyer should visit a property under consideration during different seasons. Snow especially can hide access problems such as wet areas, rock outcroppings, or impassable roads. A stream that a buyer views as an important aesthetic feature may be intermittent and may disappear in summer.

When considering the purchase of a piece of woodland, the prospective buyer should identify ownership objectives as precisely as possible. The guidelines in this chapter can be used to evaluate the potential of a parcel to meet specific goals. If substantial immediate income from a property is needed to recoup a portion of the purchase price, property with merchantable sawtimber will be required, but it should be remembered that the cutting of sawtimber will alter the appearance of the land. If immediate income is less of a need, properties dominated by pole-sized trees and small sawtimber may be more attractive, especially if the sale price of the land is correspondingly lower. Buyers should also look at patterns of land use in the immediate vicinity of the parcel that is for sale. If peace and quiet are the main ownership objectives, a buyer will avoid an area where housing developments are springing up.

Despite the analytic and objective tone of this chapter, forestry values should not override the emotional or aesthetic preference of a buyer for one piece of land over another. But forest management can maintain or improve the features that initially attracted a buyer, so it is often the best tool for realizing a particular vision of the land. Practically speaking, management also offers a means of defraying the costs of owning land. The buyer who considers forest management while looking at woodland may have a clearer dream and a better chance of realizing it.

3   Foresters

Most woodland owners will need the professional assistance of a forester in defining their management goals and implementing their decisions. The choice of a forester and the efficient use of a forester's time depend on a landowner's understanding of what foresters do, and of the types of foresters available to assist private landowners.

Foresters are registered or licenced in some New England states (see appendix 3). In other New England states, there is no control over who designates himself or herself as a forester. The Society of American Foresters, the profession's largest national organization, recognizes as a trained forester a person having at least a four-year forestry degree from an accredited institution.

A forestry school curriculum includes management planning, forest inventory, tree marking and sales, logging road and landing location, mapping and basic surveying, timber stand improvement, and woodland appraisals. Not all foresters are trained in wildlife habitat management, tax computation and planning, tree planting, forest recreation development, Christmas tree management, and sugarbush management, which are specialties.

Services

Even a landowner who is interested in actually undertaking the woodswork should hire a forester for certain crucial services: management planning, forest inventory, tree marking and sales administration, road layout, appraisals, and damage assessment.

Management Planning and Forest Inventory

Management planning claims a large part of most foresters' professional time. A management plan is a written, detailed description of a piece of woodland, a record of the purposes of management, and prescriptions for the appropriate choice, sequence, and timing of management activities, such as thinnings and harvests. A plan usually projects ten or fifteen years into the future and is updated regularly. Chapter 4 describes management plans in detail.

A timber inventory should be a reliable, statistically-based estimate of the sawtimber, pulpwood, or fuelwood volume in each stand on a property, by species and diameter of tree. It also includes estimates of the density of the trees and of the distribution of tree diameters in each stand. While inventories of commercial wood products and potential products are by far the most common, it is desireable to make note of other resources, such as shrubs, flowers and grasses, wildlife habitat elements, and evidence of wildlife use. Not all foresters are familiar with these techniques. In taking an inventory, a forester uses a systematic procedure to sample the relevant attributes of a forest. The sample data, along with field observations of access and site quality, are used in the formulation of management prescriptions.

Without a considerable investment in study, it is difficult for a nonforester to acquire the planning and inventory skills needed to manage tracts of forest land larger than 10 acres. For small holdings of 10 acres or less only a little professional assistance is usually necessary, and sufficient information may be available in literature from universities, the Extension Service, state forestry departments, and the U.S. Forest Service (see appendix 4). Investment in a forester's planning and inventory services is a necessity on larger tracts, especially where several forest types are present, or where problems of access are significant.

In some situations, what is often referred to as a "woodland exam" may be requested before or instead of a full management plan or inventory. In a woodland exam, a forester will take a quick look at a property to assess its current condition and its potential for achieving specified objectives, and will provide the landowner with a brief written report. A woodland exam would be appropriate for tracts of less than 25 acres, or when evaluating a piece of land for sale. The exam would reveal whether a more formal, time-consuming, and expensive look at the land was called for.

Tree Marking and Sales

Tree marking and sales are also frequent activities for many foresters. When marking timber, fuelwood, or pulpwood, a forester designates with paint which trees are to be cut, and estimates the volume of wood they

Fig. 27. A tree marked for harvest. Special paint and a paint gun are used for marking trees.

contain. The volume of marked trees usually is determined by measured diameter and estimated merchantable height.

Choosing which trees to mark is both an art and a science. A forester must consider a tree's present financial value as compared with its potential value if left to grow; the impact of its removal on the stand; its value as a seed source or for wildlife habitat; its vigor; where it is likely to fall if cut; and whether it can be profitably removed from the forest. The forester uses a paint gun to mark the tree both at eye level and at the stump near the ground (fig. 27). After logging, stump marks provide an assurance that only marked trees were cut.

When all trees to be sold are marked and measured, the forester creates a table that summarizes sawtimber by species, diameter, and volume, and sometimes by quality grade. In New England, sawtimber trees are most often sold woods run (ungraded), since the portion of a sale made up of high-quality logs is usually small. Information about marked pulpwood and fuelwood is usually less detailed, since these products are of lower value and species is of less importance.

When the forester involved is a private consultant acting as a landowner's agent, timber is often put up to bid. After marking the trees for sale, the forester sends a summary of the tree volumes and a prospectus outlining the conditions of sale to qualified bidders. At a specified time, the forester conducts a showing of the sale area. Interested buyers then submit sealed bids. With the advice of the forester, the owner decides which bid, if any, to accept. When the bid is awarded, a written contract, usually provided by the forester, confirms the price and spells out the conditions of sale. Usually the forester supervises the logging job in order to assure compliance with the contract. A landowner can arrange for tree marking and logging administration as separate services, however.

Putting sawtimber up to bid ensures that timber will bring its full current market value. For a variety of reasons, the sale of timber is sometimes negotiated directly with a single buyer, bypassing the bidding process. The most common reasons for negotiated sales are a logger's reputation for good work, the desire of a landowner or forester to have a particular piece of specialized logging equipment on an operation, or the involvement of an industrial, rather than a consulting, forester. An industrial forester is employed by a wood-using industry, and is interested in seeing that the wood does not go to a competitor. Relatively few fuelwood and pulpwood sales are put up to bid; the expense of the bidding process is often not justified, and the value of fuelwood and pulpwood fluctuates much less than that of sawtimber.

Logging Road and Landing Location

A forester hired to mark and sell timber usually designs and lays out forest roads (including bridges, stream crossings, and drainage structures) for transporting timber from the woods, and the landings where timber will be piled. This service is one for which a forester can be hired separately.

A forester is often involved in some or all of these aspects of road installation: road location to meet minimum environmental standards, long-term needs of forest management, and ownership objectives; specification of the type, number, and placement of erosion control structures such as waterbars and culverts; negotiation of the work to be done by the logger or an excavating contractor; and supervision of the contractor's work.

Mapping and Basic Surveying

Before any management activities can begin, and especially before any harvesting takes place, the land boundaries of a property must be located and marked (see the Boundary Maps section of chapter 4 for details). A forester will charge at an hourly rate for mapping, surveying, boundary location and marking services, which often include deed research, looking for physical evidence of old boundaries and corners, conferring with neighboring owners, and blazing trees along boundary lines. Foresters will not usually undertake boundary work for a lump sum. Owners who find research enjoyable might attempt it themselves.

Once the boundaries have been found and marked, maintaining their visibility is a straightforward process requiring only a good, longlasting paint and a hatchet. A forester can be retained to maintain boundaries, but his or her services may be unnecessarily expensive for this task.

When a management plan has been ordered, the fee for a map should be included in the fee for the plan if the forester is given a legal written deed description or a correct boundary map. If the forester has to take bearings and measure distances of boundaries in order to make a map for the plan, there will probably be an extra charge, calculated by the hour.

Timber Stand Improvement

Timber stand improvement, or TSI, is a collective term for several techniques for giving a competitive advantage to the best trees in a young stand and for improving their form. TSI includes *release cutting, improvement cutting, thinning,* and *pruning,* all of which are described in detail in chapter 5. Most TSI techniques involve the removal of unmerchantable or otherwise undesirable trees, by either cutting them down or killing them in place with a cut around the circumference of each tree's trunk that severs the tree's water and nutrient transport systems. Killing trees by this method is called *girdling.* TSI is often accomplished by girdling, and many foresters use the two terms interchangeably although their two meanings are not synonymous. TSI is a group of techniques, and girdling is one means of applying some of them. Girdling is further discussed in chapter 5.

A forester will designate an area for TSI and mark the trees to be felled or girdled. Some foresters will do the actual girdling or felling; others will locate a woodsworker to do the job. Today foresters are hired less for girdling work than they were ten years ago because some of the trees that would have been girdled then are now saleable for firewood.

The cost of a forester's services for TSI must be viewed as an investment in the future value of a timber or sugarbush stand, or as an investment with "unpriced" returns—more wildlife or more aesthetic enjoyment. Federal cost-sharing money is available in many counties to reduce the out-of-pocket expenditure the landowner must make for TSI practices.

Timber Appraisals and Damage Assessments

A forester can estimate the financial value of the timber on a tract of land by inventorying the merchantable timber and calculating its worth according to current prices. This service is usually requested by realtors, prospective land buyers or sellers, or attorneys involved in estate valuation and settlement. Sometimes the appraisal includes an estimate of the future value of young tree crops that are not yet mature. A forester will also appraise a forest's past financial value, a service usually performed for the purpose of calculating federal timber taxes. By various methods, foresters can estimate the value of timber lost in forest fires, disease outbreaks, or timber theft.

Tax Planning and Computation

Taxes should be a factor in the management planning process. Income, property, yield, and estate taxes are the principal types associated with forest ownership. A forester will not serve as an overall tax advisor, but should be able to assist to some degree with tax calculations and advice. Some can advise on the tax implications of conservation restrictions and easements and other legal techniques for assuring the perpetual conservation of land. Income from sales of sawtimber, fuelwood, or pulpwood is usually eligible for treatment as a capital gain. From the details of a forest inventory and sale data, a forester can do the calculations necessary to minimize a landowner's income tax liability. This service is important, because many accountants and Internal Revenue Service (IRS) agents have little familiarity with forestry and such related matters as the timber depletion allowance (see chapter 7). A forester can also help to time income-producing activities to minimize tax liability.

Each New England state has a property taxation program, often known as a "use-value," or "current-use" program, by which forest land can be assessed for less than its fair market value (see appendix 3). The purpose of these programs is to encourage the retention of forest land by relieving some of the financial pressure that leads to land development. A forester can prepare the plan and necessary forms required in each state.

Yield taxes are assessed in some states (see appendix 3) when products are harvested. A forester can calculate the amount of any yield tax due and provide the necessary forms for filing.

A forester's involvement in estate tax matters is usually limited to the appraisal of present or past value, as discussed above.

Tree Planting

Foresters in New England do little planting. For one thing, the relatively humid climate of the region contributes to a prolific natural regeneration of trees. Second, the rocky terrain often makes planting difficult. Finally, the valuable hardwood species native to New England do not survive planting very well. Some private foresters do plant trees, generally charging by the acre, and most will advise on planting, charging by the hour. On small acreages, planting is a job that an owner with the interest and time can easily handle. There is good printed information on planting techniques (see appendix 2).

Christmas Tree Management

A minority of foresters offers Christmas tree management services. The Christmas tree business provides an opportunity for substantial income from unused open land, but the operation must be well managed to succeed. The work is intensive and involves clearing or mowing, tilling, planting, periodic shearings, weed and insect control, and marketing. Many Christmas tree growers do this work themselves, with occasional professional advice. The subject is not treated further in this book, but some excellent technical references are available (see appendix 2).

Recreational Development

Most recreational development involves laying out and constructing trails for hiking and skiing, thinning woodlands to improve the looks of trailside areas or to create vistas, and pond construction. Some foresters provide such services; so do recreation managers, but they are usually employed in the public sector and are not commonly available for services on private woodlands. Plentiful advice on pond construction is available from the Soil Conservation Service.

Sugarbush Management

Although some foresters can provide consultation on all aspects of maple syrup production, those who list sugarbush management among their services most often are offering to assess a stand's potential for sugar production and oversee its development into a sugarbush, rather than to assist in the actual manufacture of syrup. Foresters skilled in sugarbush management are most abundant in New Hampshire and Vermont.

The processes of sugarbush management are identical in many respects to those of timber management—mapping, inventory, management planning, periodic thinnings—but the techniques are quite different. For a "hobby" sugarbush, an owner may be able to depend on advice from a forester supplemented by printed information on sugarbush management, which is available from various sources (see appendix 2). For larger sugarbush enterprises, a forester's services are advisable.

Types of Foresters

Not all foresters are competent in or interested in all of the services described above. There are four types of foresters available to assist private landowners in New England, distinguished primarily by who employs them. County, consulting, and industrial foresters provide on-the-ground management assistance. Extension foresters provide information only, except in New Hampshire.

County and Extension Foresters

Service or "county" foresters are state employees whose salaries are paid with federal and state taxes. They are not affiliated with the U.S. Forest Service, but are employed by the forestry departments of each New England state government. The exception is New Hampshire, where county foresters are employed by the Extension Service; in New Hampshire, the titles "Extension forester" and "county forester" are synonymous. In other New England states, the two are not the same: county foresters provide management advice to individual landowners; extension foresters offer publications, educational sessions, and answers to general forestry questions. Addresses for Extension foresters in each New England state are given in appendix 4.

In the New England states there is usually no more than one county forester for a county. Because of their sparse distribution and the potential for competition with private foresters, regulations limit to a few days the annual availability of a county forester to any one client.

County foresters have the same training as any other forester, including college-level courses in management planning, inventory, marking and sales, appraisals, TSI, and basic surveying. However, because of the restrictions on the time they may spend with any one landowner, and because of the large number of landowners in each county, they may refer technical management work to private foresters. For a landowner, a county forester serves most efficiently as an initial contact; the forester can provide referrals and an indication of woodland potential. Management services rendered by county foresters are most useful for tracts of 25 acres or less, given the foresters' time limitations.

County foresters are also responsible for administration of federal cost-sharing programs. These programs provide direct federal incentives for such management practices as timber stand improvement, tree planting, and erosion control on forest roads. Availability of funds varies by county and from year to year. Information about current year's funds can be obtained from the Agricultural Stabilization and Conservation Service (ASCS), part of the U.S. Department of Agriculture. See appendix 4 for ASCS addresses in each state. Final approval of eligibility for cost-sharing programs rests with the county forester.

County foresters are often the initial contacts for landowners interested in the Tree Farm program. This program, funded by forest industries through the American Forest Foundation, recognizes landowners who manage their land, but is not associated with any government agency, does not provide tax incentives or credits, and is not limited to Christmas tree plantations. The fact that many county foresters provide information about the program and perform Tree Farm certification inspections tends to confuse landowners. An owner wishing to enroll in the Tree Farm program will have his or her property inspected by a participating forester; consulting and industrial foresters, as well as county foresters, are often Tree Farm inspectors.

Consulting Foresters

Consulting foresters are usually self-employed; some work for consulting firms. They are hired by landowners and act as an owner's agent. They are paid on a fee or retainer basis for their services, charging by the acre, by the hour, or by the unit volume of wood, or they may receive a percentage of timber sale receipts (see table 7). Up-to-date ranges of costs should be available from county foresters. A listing of consulting foresters' organizations can be found in appendix 4.

For landowners, the advantages of private consultants are that they can act as an owner's agent in all phases of management, there is no restriction on the amount of time they may spend with any one client, and they do not have to consider the needs and interests of some wood-using industry in addition to those of the owner. A consultant is more likely than an industrial forester to be involved with those aspects of land management not

Table 7 Consulting Foresters' Fees in New England

Service	Fee Basis	
	Most Common	Also Used
Management planning	Per acre	Flat fee; per hour
Inventory/appraisal	Per acre	Flat fee; per hour
Timber/fuelwood/pulpwood marking and administration	Percentage of sale price; per cord or thousand board feet	Per hour
Marking only	Per hour	Per cord or thousand board feet
Timber stand improvement	Per acre	
Boundary mapping, location, blazing	Per hour	Per acre; per linear foot
Planting	Per acre; per thousand seedlings	Per hour
Pruning	Per linear foot; per acre	Per hour
Consultation	Per hour	

directly related to timber production such as habitat management or aesthetic improvements.

Although price is significant in hiring a consulting forester, good communication is the most important factor. After interviewing a few consulting foresters, an owner should choose the one who has explained most clearly the processes and details of forest management, and the one who seems most responsive to the landowner's objectives and needs. When interviewing a forester, a landowner must convey from the very beginning a clear picture of her or his woodland objectives. *Foresters are usually trained with an emphasis on timber management, and unless directed otherwise they will project that emphasis.*

References from a forester's other clients should be obtained, and where wildlife habitat management, sugarbush development, or Christmas tree growing are of interest to the owner, the forester should be asked about experience in those areas. Since not all states presently require the licensing of foresters, an inquiry about a forester's training is also strongly advised. A four-year bachelor's degree in forestry should be the minimum criterion in considering a forester's credentials. A sample management plan should be requested from each forester interviewed, and reviewed for comprehensiveness by comparing it to the management plan description in chapter 4.

It is important for an owner to decide what a forester will be hired to do.

It is not necessary to hire a forester for complete management services; the landowner sets the degree of responsibility the forester will have. A forester can act as the owner's agent in all phases of the management process, or perform only certain specified tasks, although a comprehensive, long-term owner-forester relationship is usually most satisfactory.

Consultants may charge by the hour for a short initial woodland tour to informally assess woodland potential. This service is one for which a county forester also is well suited, and there is no fee for the assessment.

To evaluate a forester's approach to timber harvesting, sites of recent timber sales in which the forester was involved should be visited. If the forester escorts an owner to the sites, there will probably be a charge for the time.

Although a consultant acts as a landowner's agent, conflict-of-interest situations are possible. For their timber marking and sales services, many consultants charge a percentage of the gross sales price, or by the amount of wood marked and sold. In some cases, these common practices may work as an incentive for a forester to mark more trees than is silviculturally desirable. If the fee is set as a percentage of the price, there may be an incentive to sell more of the high-quality timber. It is the responsibility of the landowner to learn from the forester the rationale of any sale that is proposed, and to make clear to the forester the ownership objectives.

There are a few foresters who provide management services and who are also loggers; these interests may conflict. As foresters, they should be practicing silviculture that meets the owner's objectives and works for the long-term productivity of the land. As owners' agents, foresters should be working to obtain the best possible stumpage price and to improve stand quality. A logger's interest, however, is in obtaining stumpage at a low price and in removing the most valuable trees. It is not categorically risky to deal with a forester-logger, but it is important that an owner be aware of the potential conflict of interest and assess it in the light of the particular situation.

A landowner who works with a consulting forester should have some form of written agreement covering costs, specific tasks, and the duration of the agreement. Normally, the forester will supply the contract. If the owner has any doubts about the agreement, referring it to an attorney may be worthwhile. A sample contract is reproduced below, with annotations on the content and purpose of the main clauses interspersed where appropriate.

CONTRACT WITH CONSULTING FORESTER

Agreement made this _____ day of _____ , 19 ____, between __(landowner's name)__ , hereinafter called the "owner," and
__(forester's name)__ , hereinafter called the "forester."

The forester will provide the following forest management services for the
owner at property located in _____ .

This sample contract covers the most common services provided by consulting foresters. Many others might be included. For each service, a clear
description of the scope of the work and the fee charged must be included.

1. *Boundary Location and Marking.* The forester will locate, blaze, and paint,
at approximately 30-foot intervals, the boundaries of the property.
Fee: _____/hour_____ .

The method of boundary marking is specified. When the corners and location of a line have been positively determined by a surveyor, the forester
will blaze and paint the boundary.

2. *Management Planning.* The forester will provide the owner with a written
management plan based on a field cruise. The plan will include: a map of forest
types, stand and site descriptions, a statement of timber volume and value, and
recommendations for management. Timber volume estimates will fall within a
_____ % confidence interval. Fee: _____ .

The contents of the management plan are detailed. The forester will sample variables of interest in the forest and use the data gathered to prepare
the plan. Specification of the confidence interval makes the forester ac-
countable for the accuracy of the volume figures; see chapter 4 for a discussion of the confidence interval. This clause represents an agreement
between the landowner and the forester on the type and amount of information to be contained in the management plan.

3. *Timber Marking and Sales Administration*

(A) The forester will mark with paint and tally merchantable sawtimber/fuel-
wood/pulpwood, located and described as follows: ____(tree quality,____
species, and size) . Sawtimber will be tallied on a 100% basis; pulpwood
and/or fuelwood will be tallied as follows: __(percentage or sampling__
method described) .

The forester marks the trees to be harvested with paint. Where all trees in a
given area are to be cut and are of low value, the forester may just designate
the boundaries of the area. In a 100 percent tally (normal in valuable
timber), each tree is measured for diameter and merchantable height is estimated. With low-value material, such as fuelwood and pulpwood, mea-

surement may be taken from a sample of the marked trees; with small, low-value material, the time required to measure the diameter of every tree may not be justified.

When the cutting of pulpwood or fuelwood is to be heavy, a forester will sometimes mark only the trees to be *left,* thereby saving considerable time. The contract should indicate if this is to be done.

(B) The forester will provide owner with a written tally of the marked timber.

The owner receives a copy of the tally. Normally, a tally of sawtimber will show the number of trees in different species and diameter groups, and the volume present in each group. If the timber to be sold is graded according to its quality, the tally will also categorize trees by grade as well as species and diameter. Tree grades are determined by their diameter, merchantable length, straightness, and the number of visible defects, such as scars and seams. Graded tallies are uncommon in New England.

(C) According to the owner's wishes, the forester will solicit sealed bids for or negotiate a sale of the marked timber/fuelwood/pulpwood.

The forester should discuss with the owner the relative merits of negotiated and bid sales (see chapter 6). The owner should decide which route to take. Timber sales should be advertised only to reputable loggers.

(D) The forester will advise the owner as to the acceptance or rejection of bids and offers received.

The forester will be able to evaluate any offers or bids received, and to advise the owner on the possibility of getting higher bids by waiting a while. Timber markets fluctuate quite a bit. A forester should be able to give an owner an opinion concerning when and to what extent a given market might improve. With this information, an owner can evaluate a current offer.

(E) If the owner accepts an offer or bid, the forester will prepare a logging contract suitable to both the owner and the buyer.

(F) The forester will collect and distribute sale revenues as specified in the contract signed by the owner and the buyer.

The responsibility for collecting sale revenues is normally part of the forester's job. The check for sale proceeds can be made payable to the forester, who will deduct any fee that is due and remit the balance to the owner. The landowner may collect the proceeds and pay the forester.

(G) The forester will collect, hold in escrow, and distribute any performance monies involved in the sale.

Most consulting foresters maintain escrow accounts to hold performance deposits during logging jobs. A performance deposit is a sum of money, independent of and in addition to stumpage payments, provided by the buyer and held by the owner or forester while logging is underway. It is meant to cover the costs of stolen trees, excessive damage, or road repair work resulting from logging. It is used by the forester to enforce contract compliance, and it is returned following completion of a logging job. An alternative is the posting of a bond, which a logger can purchase from an insurance agent.

(H) The forester will supervise the harvesting job by making periodic inspections to insure successful execution of the logging contract.

Inspections should be as frequent as necessary. This is probably the most critical part of the forester's job during a timber sale, and it is very time consuming. The landowner depends on the forester's professional interpretation of key provisions of the logging agreement.

(I) On behalf of the owner the forester will file any forms required by state, federal, or local laws and bylaws relating to the sale, except income and yield tax forms.

Some states have filing requirements when a timber/pulpwood/fuelwood sale is made (see appendix 3).

(J) The forester will lay out roads and landings necessary for the sale, subject to the owner's approval.

(K) In consideration of these services, the owner will pay to the forester the following fees: _____ .

Foresters charge in one of several ways for timber marking, sales, and administration, as shown in table 7. Some foresters charge separately for marking and administration. A forester may charge by the hour for inspections of logging jobs, but this practice is not common.

(L) If for any reason the owner decides not to sell the marked trees, the owner shall pay to the forester _____ per thousand board feet of marked timber, or _____ per cord of marked fuelwood/pulpwood, as applicable, in payment for services performed, and shall also pay to the forester the reasonable costs incurred by the forester in soliciting bids and/or negotiating the sale of timber/pulpwood/fuelwood.

The forester requires some protection to insure that he or she gets paid for work already done if the owner decides not to sell the timber after it has been marked or put up for sale.

Terms and Conditions

1. The owner warrants that he/she is the true owner of and has the full legal right to mark and/or sell any timber/pulpwood/fuelwood to be marked or sold under this agreement, and that there are no other claims to said timber/pulpwood/fuelwood; or that the undersigned is the owner's agent in this matter and has the authority to mark and/or sell the same.

This clause protects the forester if any question arises regarding the owner's legal right to dispose of any timber cut, or regarding the status of the forester as the landowner's agent.

2. The owner agrees to provide the forester with accurate information regarding the location of the boundaries of the property, and if any boundary is in dispute, to so indicate to the forester. The owner agrees to indemnify and hold harmless the forester from any claim for damages to the trees and/or property of abuttor(s) as a result of owner's incorrect representation of boundaries.

This clause obligates the owner to provide the forester with correct information on property boundaries. If trees are cut on a neighbor's land, the forester is liable only if negligent, and not if the owner provided poor information.

3. The owner agrees to permit access to the property by the forester at all reasonable times.

4. All contract(s) for the sale of timber/pulpwood/fuelwood shall be signed by the owner. The owner assumes full responsibility for his/her performance under said signed contract(s).

Although the forester is the owner's agent in any sale of standing trees, the owner is the signatory to the sales contract.

5. The owner agrees to indemnify and hold harmless the forester from liability for personal injury or property damage arising out of this agreement, and/or out of any contract for the sale of timber/pulpwood/firewood unless said injury and/or damage is caused by the willful and reckless act(s) of the forester.

This clause insures that the forester is not liable for a logging injury or accident while the forester is acting as the landowner's agent, unless the forester acts intentionally, recklessly, or negligently to cause the injury or accident.

6. This agreement shall be binding on all parties hereto, and shall expire on ___(date)___ , or upon successful completion or expiration of the timber/pulpwood/fuelwood sale contract. If the owner decides not to sell the marked timber/pulpwood/fuelwood, this agreement shall expire upon payment of any fees that are due.

A date certain serves to end the agreement in the case that no buyer is found.

Industrial Foresters

Industrial foresters are employed by sawmills, paper mills, and other major wood-using facilities. Procurement of a constant supply of raw materials for the employer is the industrial forester's primary concern. In a few localities in New England wood-using industries agree to provide a forester's services to private landowners in exchange for long-term cutting rights or the right of first refusal on saleable timber. Under such an agreement, the company may be entitled to purchase any timber the landowner chooses to sell if the company matches any offer made to the landowner in good faith by another buyer.

Because the industrial forester's job is to ensure a reliable flow of fiber to the employer, the forester is likely to provide a private landowner with timber management only, and may not offer such services as wildlife, aesthetic, or recreational management. Industrial foresters are first and foremost timber managers and buyers (although *not all timber buyers are foresters*), and they must answer first to their employers.

Industrial foresters usually initiate a relationship with a landowner, but an owner interested in the services of an industrial forester might make contact through the forester's employer. A list of wood-using industries is usually available from the Extension Service. There is often no charge to the landowner for the services of an industrial forester, who may operate in one of several ways. A forester may mark and tally the trees to be cut, and the owner is paid by the company on the basis of that tally. In other cases, the forester may only mark the trees, and the owner is paid according to the amount of timber that arrives at the mill. A third method involves no marking, since cutting is guided by a "diameter limit," as discussed in chapter 5.

The principal advantage to an owner in contracting with an industrial forester is the reliability of an outlet for timber and the lack of a fee. Services are "free," however, only if the timber is sold for what it is worth and the forest is left in good condition.

In some situations, an industrial forester may be faced with a conflict of interest. As the landowner's forester, he or she should recommend and implement management activities that best serve the needs of the land and landowner. As an employee of a wood-consuming industry, the forester's goal is to procure a supply of high-quality material for the company at as low a price as possible. Aside from the question of price, the quality of the work is important: is the company willing to take the low-value trees that should be cut, as well as the more valuable, in order to improve the stand?

A sample contract with an industrial forester is reproduced below, with appropriate annotations.

CONTRACT WITH AN INDUSTRIAL FORESTER

Right of First Refusal to Purchase Timber
Agreement made this _____ day of _____ , 19 _____ , between ___(landowner's name)___ , hereinafter called the "owner," and ___(company name)___ , hereinafter called the "company." This agreement shall bind the heirs, successors, and assignees of both signatories.

Agreements with an industrial forester may take several forms. The most common is based on the "right of first refusal."

Terms and Conditions
1. The owner grants to company the right to purchase timber on the owner's property, as located and described below: _____

_____

2. If the owner decides to sell any or all of the timber located on the above-described property for a period of _____ years following the signing of this agreement, the owner grants to the company the right of first refusal on the timber being sold.

These clauses establish the principle of the right of first refusal. Commonly, such contracts are written to cover periods of five to twenty-five years.

3. If the owner receives a legitimate written offer from another buyer to purchase the timber, and if the offer is acceptable to the owner, the company shall have _____ days to match the offer. A copy of the offer will be made available to the company.

This clause insures the company of a chance to obtain timber at no higher price than other buyers are willing to pay at the time of sale.

4. If the company decides to match the competing offer, the owner and the company shall negotiate and sign a sale contract acceptable to both parties within _____ days of the owner's receipt of the competing offer.

The clause makes certain that the company enters into a contract acceptable to the landowner within a reasonable time following receipt of a competing offer.

5. If the owner has not received a matching written offer from the company within the specified time period, the owner shall be free to accept the competing offer. All subsequent offers shall be subject to the conditions of this agreement.

6. In exchange for the right of first refusal described in this agreement, the company agrees to provide at no charge, and subject to the approval of the owner, the following forest management services during the period covered by this agreement: _____

_____
_____
_____
_____
_____

Typical services provided under this clause would be boundary maintenance, management planning, property taxation assistance, and TSI activities covered under various government cost-sharing programs. The landowner receives copies of all documents and records prepared in relation to these services.

7. The company and the owner also agree to the following special provisions:

_____
_____
_____
_____

Inserted here would be anything of special concern to either party, as well as charges for management activities that the company does not provide free.

Not all contracts with industrial buyers are of a long-term nature. A single timber sale may be the extent of the relationship; a thorough contract should regulate the sale (a sample sale contract is provided in chapter 6).

Landowners often sell timber or pulpwood to mills that agree to provide supervision of the logging job. Not all logging supervisors are foresters. If the buyer provides supervision, it is important that the industrial forester's inspections are frequent enough. The forester/supervisor is working for the buyer and not directly for the landowner. As a rule of thumb, inspections should be made at least weekly by the industrial forester and frequently by the owner. The owner should discuss any issues of concern with the individual in charge as soon as they arise.

It is difficult to compare the costs of a consulting forester with the apparently free services offered by industrial foresters. The services received from the industry must be evaluated in the light of the contract terms: if the industry or mill offers a written contract, does it stipulate a constant price that will be paid for wood? If it does, the fairness of the price should be investigated. What is the duration of the agreement? Under what conditions can the contract be amended, revised, or nullified? Does the agreement include commitment to an exclusive association with the company? What assurances are given that proper silviculture will be practiced?

If a landowner chooses an industrial forester, the marking of trees before a sale should be a free service. When operating under a right-of-first-refusal contract, the method of marking the trees to be harvested must be defined and all parties should be bidding on the same trees. A landowner should reserve the right to choose a method of marking other than the diameter-limit system (discussed in chapter 5), which is often undesirable.

Foresters should be more than managers and brokers of wood products. They should be able to balance the biological and economic aspects of forest management, while operating within a framework of goals and objectives provided by the landowner. Early dialogue between landowner and forester is critical to a successful relationship. A landowner must accept the role of primary decisionmaker while permitting the forester to evaluate the technical feasibility of a given objective, advise the owner, and implement decisions.

Whether to select a consulting or an industrial forester depends on a variety of factors, the most obvious of which are price, availability and good communication. Equally as important are the owner's evaluation of a forester's technical competence, the forester's willingness to recognize the full stated range of ownership objectives, and the satisfactory resolution of any potential conflicts of interest.

4 Management Plans

Managing woodland requires careful timing and sequencing of woods-work over a long period of time. A written management plan, updated regularly, allows for an orderly program of management over the life-span of trees, and provides an element of continuity as ownerships or foresters change. Possession of a written management plan is necessary for valid and thorough forest management, and, in most New England states, a plan is required to qualify for property tax relief programs for forest land.

There is no standard management plan format. In composing a management plan, a forester generally uses an outline he or she has developed. However, standardized management plan forms are required by some New England states for land that is included in their current-use taxation programs (see appendix 3). Although formats vary, any serviceable plan will include these elements:

1. Management objectives for the entire property
2. Location, boundary, and stand maps
3. Management goals, by stand
4. Stand descriptions
5. Site descriptions, by stand
6. Recommended treatments, by stand
7. Procedures for updating the plan

The first two elements in this list pertain to the entire woodland tract. Items three through six relate to individual stands or habitat management areas. The last item refers to a brief note on the frequency with which the plan should be updated. In the pages that follow each element in a management plan is discussed individually, and for each element either written extracts from a plan or illustrations are provided.

Management Objectives for a Property

It is important that the purposes of management for an entire property be stated as clearly and specifically as possible. The process of articulating, adjusting, and writing down management objectives requires the forester and the owner to share each other's perspectives on the management of the woodland. The process usually results in practical goals that are well understood by both parties, for example:

Management Objectives

The owners' objectives in the management of their woodland are as follows, in order of importance to them:
1. Recreation, nature study, and watching songbirds
2. Maintaining a source of fuelwood (10 cords per year)
3. Income from timber production of sawlogs
4. Enhancement of value for heirs
5. Sugarbush development (goal of 1,600 taps)

A statement of purpose provides important information for future owners of a property. Omission of such a statement is a common fault of management plans.

Location, Boundary, and Stand Maps

Accurate maps of the land are important elements of a management plan. A location map identifies a property on a broad-based map, such as a U.S. Geological Survey topographical map or a state highway map (fig. 28). A boundary and stand map shows the compass bearings and lengths of the property boundaries and is drawn to scale; that is, every inch on the map represents a fixed distance on the ground (fig. 29). Such a scale permits an accurate determination of acreage. It is not uncommon for deeds or maps (especially old ones) to contain inaccurate acreage figures.

A boundary and stand map should show where boundaries follow stone walls, fences, roads, and streams. The map should also identify the methods used to mark the corners. Unless corners and boundaries are marked, the landowner runs the risk of *timber trespass*—of harvesting someone else's timber, or having his or her own trees taken by mistake. (The usual penalty for timber trespass is an award to the plaintiff of one to five times the current market value of the trees, plus damages.) In Maine, it is required by law that corners and property lines be established before timber can be harvested. While only a registered surveyor can legally locate a new corner or boundary, anyone can relocate and mark an existing line. The services of a registered surveyor are expensive, but they should be used when a deed is rewritten to more accurately describe boundaries, when a map is to be legally registered, when there is a serious dispute or question

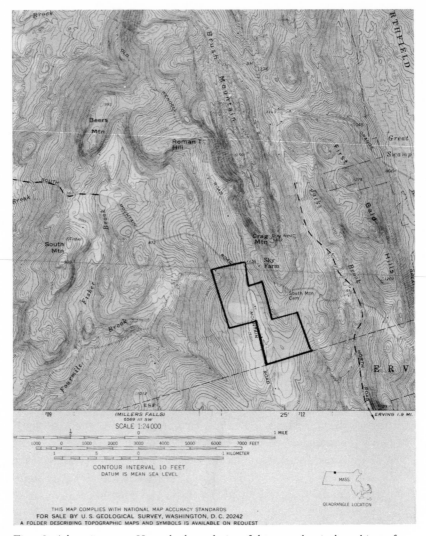

Fig. 28. A location map. Here, the boundaries of the tract that is the subject of the management plan are drawn on a USGS topographical map.

about a boundary or corner, and when future subdivision of the property is intended.

The job of the surveyor involves weighing all available written and physical evidence to come to a conclusion about the location of a line or corner. Surveyors sometimes serve as arbitrators between parties who disagree about a boundary, or who are trying to establish a compromise boundary from insufficient information. It is important to realize that a surveyor usually sets or resets corners, and establishes the compass bearings of the lines running between the corners. The surveyor usually does *not* mark the boundary lines connecting the corners. A forester can mark boundaries once they are located. Some foresters are registered surveyors.

Corners in New England are marked in a variety of ways. Some of the most common traditional markers include iron pipes (abbreviated I.P. on survey maps), gun barrels (G.B.), concrete bounds (C.B.), stake and stones (S&S), piles of stones, granite posts, drill holes in stones, and stone wall intersections. So-called "witness trees" are used to point to the location of a

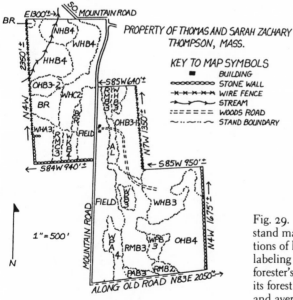

Fig. 29. A boundary and stand map. The combinations of letters and numbers labeling each stand are the forester's system of indicating its forest type, tree density, and average tree height. Foresters use many variations of letter-number coding systems.

Fig. 30. An overgrown blaze.

Fig. 31. A blaze consisting of a notch made with an axe or hatchet, then painted. Red paint is commonly used for boundaries.

corner. Witness trees usually have two or three blazes to distinguish them from boundary line trees that are not near the corner. Witness trees are especially helpful when corner markers have rotted, been buried, or been moved.

Boundary lines should not be blazed until the owner or forester is certain that the line has been located correctly. Preliminary marking can be done with flagging or strips of cloth. Locating old boundaries that have not been maintained is made easier when evidence of the old line can be found in the woods. Typical indicators of boundaries include stone walls, wire fences, remnants of wooden fences, old blazes, natural boundaries, and a change in forest type or tree size. Wire fences have often been strung from tree to tree, rather than straight along a boundary, and wire is often found embedded in trees; a metal detector can be used to locate wire that may have gotten buried. Virgina rail fences were once common in some parts of New England, and sometimes old rails or rocks that were used as a base can be located. Old blazes which have healed have a characteristic appearance (fig. 30). When reblazing a tree on which an old blaze has been found, a landowner or forester should leave the old blaze intact; it constitutes physical evidence that somebody in the past treated it as a boundary tree. In the event of a boundary dispute this evidence could be important. A new blaze should be added above or below the old one. Natural boundaries such as streams, roads, and ridge lines have often been used as property boundaries. A sudden and clear change in the species composition or size of a forest usually indicates that adjacent areas have had a different history of use, and may be evidence of a property line.

All boundary lines should be blazed and painted (fig. 31). Some owners fail to blaze lines that coincide with stone walls. This is a mistake, since many stone walls in New England are internal to a property. Boundary line stone walls should be designated by blazed lines. The purpose of blazing a line is to indicate to someone walking through the woods the location of a boundary. To be effective, blazes should be made every 20 to 30 feet along the line, and should be visible to someone coming from either direction. Few trees will be situated exactly on a boundary; most blazed trees will be located on either side of the actual line. Blazed trees should be within 10 feet or so of the line, and the blaze should face the line. The occasional tree that does fall exactly on a line should be blazed on the two sides facing the line.

A traditional blaze is made by chopping out a 4- to 6-inch square sec-

Table 8 Principal Forest Types of New England

Type	Major Species	Common Associates	
Spruce/fir	Balsam fir Red spruce	White spruce White pine Black spruce Larch Northern White cedar Hemlock	Red maple Yellow birch Paper birch White ash Beech Sugar maple Trembling aspen
Northern hardwoods	Sugar maple Beech Yellow birch Red maple Paper birch	Red spruce White spruce Hemlock White pine	Basswood Black cherry Elm White ash
Transition hardwoods	Red oak Black birch Red maple Paper birch	White pine Hemlock Hickories Black oak White oak Chestnut oak	Yellow birch Beech Sugar maple Black cherry Elm
Central hardwoods	Red oak Yellow poplar Hickories White oak Black birch Red maple	White pine Hemlock Black oak Scarlet oak Chestnut oak	Sugar maple Beech White ash Elm
White pine	White pine	Any of the major species or associates of central or northern hardwoods, depending on location within the region	
Hemlock	Hemlock	Any of the major species or associates of central or northern hardwoods, depending on location within the region	
Pitch pine/oak	Pitch pine Scrub oak	White pine	

Source: F. H. Eyre, ed., *Forest Cover Types of the United States and Canada* (Washington, D.C.: Society of American Foresters, 1980).
Note: The list is not all-inclusive; pioneer species are omitted. Stand composition varies by site, land-use history, and successional status. Forest types overlap in many areas.

tion of bark down to live wood. After the wood dries, the blaze should be painted a bright color with a long-lasting paint. Trees to be blazed should be vigorous and at least 4 inches in diameter. Blazes should be made at eye level or higher for easy visibility, especially in those parts of New England which receive a lot of snow. Some owners may find the traditional blaze objectionable, since it involves injury to a living part of the tree. An alternative is to use a drawknife to smooth the bark without cutting into the wood. A blaze of this kind has to be renewed more frequently. Boundary lines can be further highlighted by *brushing out,* which involves cutting out very small trees and shrubs and removing low-hanging branches on the line so one can sight along an unobstructed corridor running the length of the line.

A boundary and stand map should also show the division of a property by forest *stands,* displaying the land area covered by each stand. A stand is a contiguous area where the species, size, age, and general condition of the trees is uniform enough to be distinguished from adjacent areas. A stand often coincides with a *forest type.* Forest types result from the fact that, with some predictability, several species tend to grow in association with one another. Many factors determine the mix of species occupying a given area, including latitude, elevation, aspect, soil type, land-use history, and successional status. Table 8 lists the principal forest types occurring in New England, as well as commonly associated species; the map in figure 32 indicates where different forest types predominate in the region. Distinguishing forest types in some areas of New England may be difficult, since stands often overlap and mix. Although the edges of stands appear definite on a boundary and stand map, they are often not as obvious in the woods. Where one stand changes gradually into another, the dividing line is a matter of judgment.

In addition to boundaries and forest types, a boundary and stand map should show topographic features such as hills, streams, ponds, swamps, ledges, and roads. These are important in designing access to the land, assessing potential wildlife habitat, and locating recreational and aesthetic areas.

Stand Management Goals

In a management plan, management goals, stand and site descriptions, and recommendations for treatment of stands are given for each stand, one

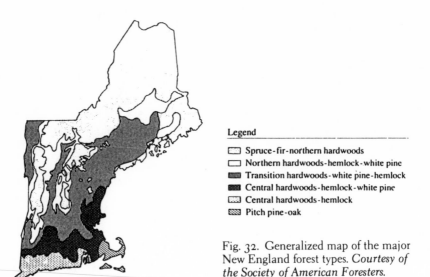

Legend

☐ Spruce-fir-northern hardwoods
☐ Northern hardwoods-hemlock-white pine
▨ Transition hardwoods-white pine-hemlock
▨ Central hardwoods-hemlock-white pine
▨ Central hardwoods-hemlock
▨ Pitch pine-oak

Fig. 32. Generalized map of the major New England forest types. *Courtesy of the Society of American Foresters.*

stand at a time. In this section and in the two following sections of the chapter these elements of a management plan are explained by discussions of two sample stands—Stand 101 and Stand 102.

A management goal for a stand is specific and is developed from the general purposes of management described at the beginning of the plan.

Stand Management Goals

Stand 101
Improvement and diversification of songbird habitat, and production of high-quality sawlogs, while safeguarding aesthetics.

Stand 102
Production of high-quality hardwood sawlogs, with fuelwood as a by-product of early thinnings. Safeguard wildlife habitat and water quality.

Stand Descriptions

In a stand description the important vegetational characteristics of each stand are given. Elements of the description are usually of the overstory trees only and include tree species, age, diameter, stand density, volume, quality, and growth rate. However, it is also essential that any important wetlands, rare or endangered species or habitat, and any other fragile areas are also noted. The forester bases a stand description on information gathered from sample trees during a forest inventory. Because of the amount of work that would be involved in measuring or assessing every tree in a forest, a forester examines a small sample of the total number of trees. The number of places where sampling is done depends on the degree of accuracy desired and on the diversity of the forest: the more heterogeneous the woodland in terms of tree species, age, diameter, density, quality, and growth rate, the larger the number of sample points.

When the size of the sample that is needed has been determined by the forester, the sampling points are distributed throughout the property in such a way as to give the most important, valuable, or least uniform forest types more representation in the inventory. Information on timber quality and on tree species, age, diameter, crowdedness, and growth rate that is gathered at the sampling points in a stand is assumed to be typical of the entire stand, and is expanded mathematically to provide an estimate of the characteristics of the whole. If an owner's interest is wildlife habitat management or botanical study, shrubs and wildflowers are included in the inventory. A summary of the stand inventory is provided in a table that is part of the stand description.

Stand Description

Stand 101
Type: White pine—Hemlock—Red maple
Acreage: 15
Principal species in this stand are red maple, white pine, and hemlock. Respectively, these species comprise 30%, 40%, and 10% of the basal area. Black birch, white birch, and black cherry are minor associates. This is a two-storied stand. Large white pine and hemlock trees, left from a harvesting operation 15-20 years ago, are scattered through the stand. Most of these trees are from 50 to 70 years old, and range in diameter from 14 to 20 inches. The lower story, primarily red maple poles with an average diameter of 7 inches, has a basal area of 30 square feet. Total basal area per acre is 100 square feet. Overall tree form and quality is fair. Some of the white pine is of poor quality as a result of white pine weevil damage and large lower branches. These trees developed in full sunlight when the land reverted from pasture to forest. Although red maple vigor is high, this is a low-value species for sawtimber production. White pine growth rate is 8 rings per inch (1 inch increase in diameter every 4 years). Red maple growth rate is 10 rings per inch, or 1 inch in diameter every 5 years.

Merchantable Volume and Value Summary

Stand 101	Per Acre				Stand			
Species	Board Ft.	$	Cords	$	Board Ft.	$	Cords	$
White pine	5,600	336	2	—	84,000	5,040	—	—
Hemlock	1,100	33	—	—	16,500	495	—	—
All hardwoods	700	35	6	36	10,500	525	—	540
Totals	7,400	404	8	36	111,000	6,060	—	540
			$440/acre				$6,600 total value	

Confidence interval 15% (see page 99)

Stand 102
Type: Oak-hardwood
Acreage: 22
Principal species in this stand are red oak, white oak, and red maple, comprising 50%, 30%, and 11% of the basal area, respectively. Average stand age is 45 years; average tree diameter is 7 inches. Total basal area is 100 square feet per acre, higher than recommended for sawlog production. Except for red maple, tree quality is good. The stand has sustained moderate defoilation by the gypsy moth, but there has been no mortality, and current egg mass numbers are moderate. Red oak is the favored crop-tree species in this stand. Current growth rate of red oak is 9 rings per inch, or 1 inch in diameter growth every 4.5 years.

Merchantable Volume and Value Summary

Stand 102	Per Acre				Stand			
Species	Board Ft.	$	Cords	$	Board Ft.	$	Cords	$
Red oak	800	160	9	54	17,600	3,520	148	1,188
White oak	350	35	4	24	7,700	770	88	528
Red maple	125	4	4	24	2,750	88	88	528
Other hardwoods	—	—	1	6	—	—	22	132
Hemlock	300	9	—	—	6,600	198	—	—
Totals	1,575	208	18	108	34,650	4,576	396	2,376
			$316/acre				$6,952 total value	

Confidence interval 18% (see page 99)

In stand descriptions, stand age is given as an estimate of the age of the trees in the main canopy. A stand is classified as *even-aged* if all of its trees are of similar age; technically, this means that the difference in age between the oldest and youngest trees does not exceed 20 percent of the estimated age at which the stand will be ready for harvest. For even-aged stands, stand age is given as the mean age of the trees. If a stand is composed of trees of at least three different age classes, it is designated as an *uneven-aged* stand. In this case, a range of ages best describes the stand. Knowing whether a stand is even- or uneven-aged is especially important when choosing appropriate management techniques (discussed in chapter 5).

Stand age is important information. Trees which have been suppressed in the understory of an overcrowded stand may increase their growth rate after a thinning if they are young, but may continue to decline if they are too old, despite release. For most New England species on sites with average potential, the critical age after which release may not be effective is between 40 and 70 years.

Trees must be sexually mature in order to produce seed, and their ages indicate their proximity to that condition (table 9). Seed production affects prospects for stand regeneration and the available food supply for wildlife.

The diameter of a tree is the single most important factor in its utility for a given wood product, and the single most important factor in sawlog value. The diameter of a sugar maple tree determines how many taps it can accommodate and therefore how much sap it can produce, so diameter distribution is the basis for estimating the potential number of taps in a sugarbush. Mean diameter is used to classify forest stands:

Seedling and sapling	less than 4.5 inches DBH
Poletimber	4.6 to 9.5 inches DBH
Small sawtimber	9.6 to 14.5 inches DBH
Large sawtimber	14.6 inches and greater

For uneven-aged stands, which are the exception in New England, mean diameter is unimportant; rather, it is the estimated distribution of diameters—the number of trees in the stand which are seedlings, poles, and large or small sawtimber—that is significant.

The density, or crowdedness, of a stand is commonly expressed in terms of *basal area per acre*. Basal area per acre is an estimate of the cross-sectional areas of trees on an acre at 4.5 feet above the ground (fig. 33). Ba-

Table 9 Minimum Seed-Bearing Age (Years) of Selected Trees

Younger than 10	10–19	20–29
Red maple	Paper birch	Balsam fir
Northern white cedar	Ash	Mockernut hickory
Black spruce	Black walnut	Hophornbeam
Pitch pine	Yellow poplar	Red pine
Aspen	White pine	All oaks
Cottonwood	Black cherry	North white cedar
Black locust	Basswood	Hemlock
	Elm	

30–39	40 or older
Sugar maple	Yellow birch
Pignut hickory	Black birch
Bitternut hickory	Shagbark hickory
White spruce	Beech
	Larch
	Red spruce

Source: U.S.D.A. Forest Service, *Seeds of Woody Plants of the United States,* Agriculture Handbook no. 450 (Washington, D.C.: U.S. Department of Agriculture, 1974).

Note: These are minimum ages. Considerable variation can be expected as a result of differences in genetic origin and environmental conditions.

Fig. 33. The concept of basal area. The basal area of a tree can be thought of as the surface area of the top of its stump were it cut off at 4.5 feet above the ground. The basal area of a stand is given as the sum of all the basal areas of the trees on an acre.

Fig. 34. The concept of crown cover percent. In the drawing, the crowns of the trees cover 65 percent of the acre on which the trees stand.

sal area of a tree can be visualized as the surface area of a stump were the tree to be cut off at 4.5 feet above the ground. The total surface area of all the stump tops on an acre would be the basal area per acre. Although the concept may seem strange and impractical, it is in fact clearer and less complicated than alternative methods of judging density. Knowing only the number of trees per acre is not helpful to understanding the crowdedness of a stand; a stand of large trees is more crowded than one with the same number of trees of smaller diameter. Stand basal area, accompanied by an estimate of stand diameter, more succinctly describes tree density. In timber management, basal area is useful information because it can be translated into a general estimate of the volume of standing timber.

An alternative measure of density, *crown cover percent,* is an estimate of the ground area directly under the tree canopies in a stand—the proportion of forest land area covered by tree crowns (fig. 34). Like basal area, crown cover percent is a concept of tree density which dispenses with tree numbers. It gives a better indication of light levels beneath the main canopy than basal area per acre. Light reaching the forest floor is an especially critical factor in seed germination and seedling growth, and in maintaining ground-level food and cover for wildlife. Crown cover percent is the measure of density most appropriate for planning and executing shelterwood harvests (see chap. 5), although it is less useful than basal area for estimating timber volume.

Of the two measures of density, basal area is much more commonly used in New England because a forester can relate it directly to the timber volume in a stand, and because it is easier to estimate.

Table 10 Ranges of Basal Areas for Full Stocking for Various Uses

| Average Stand Diameter (DBH in inches) | Basal Area (square feet per acre) | | |
	Northern Hardwood Sawtimber	Spruce/Fir Sawtimber	Sugarbush
5–9	55–75	70–120	20–40
10–13	75–90	120–170	40–50
14 and larger	90–95	170–200	55–60

Sources: R. M. Frank, J. C. Bjorkbom, *A Silvicultural Guide for Spruce-Fir in the Northeast,* General Technical Report NE-6 (Upper Darby, Pa.: U.S.D.A. Forest Service, 1973); K. F. Lancaster et al., *A Silvicultural Guide for Developing a Sugarbush,* Research Paper NE-286 (Upper Darby, Pa.: U.S.D.A. Forest Service, 1974); Debald, *Silvicultural Guide for Nothern Forest Types in the Northeast,* revised (Broomall, Pa.: U.S.D.A. Forest Service, 1987).

While basal area and crown cover percent are estimates of density, *stocking* refers to whether that density is too high (overstocked), too low (understocked), or within the optimal range (well-stocked) for the management goal. Stocking is a comparison of the estimated stand density (usually expressed as basal area) with some ideal density, which will vary with the management objective (see table 10). Because maple trees managed for sap production should have wide, deep crowns, sugarbushes are maintained at very low basal areas (densities). Softwoods for sawtimber production grow well at higher denisites than hardwoods, primarily because of tree form. In the case of softwood pulpwood, optimum stocking can be even higher, since diameter growth is less important than it is for sawtimber.

Optimum stocking also varies with average stand diameter. As trees grow in diameter, the number of trees per acre decreases, but the ideal basal area increases (see fig. 35).

Stand volume is the standard means of expressing the quantity of merchantable wood fiber on an acre of woodland, and it is only significant for the measurement of wood products. Volume can be expressed in board feet, cords, cubic feet, or by weight. It is estimated from diameters and heights of trees sampled during the inventory process.

The volume of trees that are large enough and are otherwise qualified to be classified as sawtimber is given in board feet. Cord measurement is used for trees that are too small or defective for sawtimber and, where legal, it is the standard measurement for firewood and pulpwood. Wood volume measurements expressed in cubic feet, cunits, and pounds, do not imply a specific product. (Units of measure are defined in appendix 1). However, cubic-foot volume measures and weight are little used largely because of the long-standing tradition of measuring in board feet and in cords in the U.S.

A summary of merchantable volume includes all material that is currently saleable. It is not an estimate of what should be cut at the present time, unless a clearcut is recommended. Value estimates are based on a variety of factors, including quality, species, access, existence of local markets, and current market conditions.

A *confidence interval* statement may accompany an estimate of timber volume. It is a statement of the reliability of the forester's estimate, and it is calculated statistically. It is equivalent to saying, "The average volume per acre is _____ board feet, give or take _____ percent." Low-value stands are usually sampled lightly, so the confidence interval percent for such stands will be high. To increase confidence in the estimate, high-value stands should be sampled more heavily. Because a more reliable estimate requires more sampling, lowering the confidence interval can be

Fig. 35. As trees grow in diameter, there will be fewer trees per acre, but ideal basal area increases. *Left:* a managed stand of white pine timber with an average diameter of 6 inches DBH might include about 600 trees per acre and have a basal area of 125 square feet per acre. *Center:* at 12 inches DBH, full stocking for the same stand could mean about 250 trees per acre and 200 square feet of basal area per acre. *Right:* when the trees reach 18 inches DBH, there would be only about 125 trees per acre but about 225 square feet of basal area per acre.

expensive. Whether it is desirable in a given case depends on the size of the property, the value of the timber, and what the information will be used for. Relatively precise inventories are justified on larger properties having valuable timber where the financial aspects of forest management are paramount.

The amount of new wood that a forest produces during a given period of time is typically expressed in a management plan as average diameter growth, or as stand volume growth. By using an increment borer, a forester can determine how many years it takes a tree in a stand to grow an inch in diameter. Determination of average diameter growth is useful because it allows the forester to project how long it will take for trees to reach sawlog size, or to become large enough to be tapped for sap. Volume growth can be calculated from diameter growth. Comparison of the volume growth rates of similar stands is the basis for allocating investments in timber management; vigorous stands and slow-growing young stands on good sites deserve the most attention because they have the potential for the highest financial return. Volume growth can be multiplied by the average current board-foot, cubic-foot, or per-cord value to determine the rate at which the stand's value is increasing.

Computer programs that use mathematical models to simulate stand growth are used by some foresters. Based on biological and economic assumptions, these programs are useful for planning purposes, although their reliability declines beyond a 20-year projection into the future.

For sugarbush management, volume growth is not important. Diameter growth is used to determine how many new taps can be anticipated each year, and therefore helps to predict future sap production.

Site Descriptions

A description of the site for each stand includes the stand's aspect, degree of slope, and *site index,* and the location of such important land features as streams, swamps, rock outcroppings, surface boulders, and other features affecting access to the stand and the quality of the site for a desired use. Present or future vistas may be noted. The location and quality of existing roads are described.

Site Description

Stand 101
Stand 101 is located just east and south of the middle field. The land is generally flat. Access to the area is good on existing roads. The soils are of medium fertility for white pine. (Site index 60 for white pine.)

Site Description

Stand 102

Stand 102 is located in the eastern portion of the property, just east of the stone wall. The land faces east, with an average slope of 20%. Access to the area is good from existing roads; however, there is no road through the stand. Wood removal will require uphill skidding. A small stream borders the stand on the east. The soils are capable of average production. (Site index 55 for red oak.)

A *site index* is an indicator of the quality of a site for growing trees. The site index of a stand is expressed as a single number that represents an estimate of its average height in feet at a given age. Height is used as an indicator of site quality and stand vigor because it is a feature little affected by other conditions, such as the crowdedness of trees (see chap. 2). Height growth, and therefore site index, cannot be significantly improved through forest management. The site index is an estimate of the inherent capability of the land to produce wood; full capacity can be realized, but it cannot be improved unless such impractical measures as fertilization are undertaken.

By convention in New England, the given age used in reference to site indexing is 50 years: site index is a rating of how tall a stand was or will be at age 50 (fig. 36). The taller the trees at age 50, the better the site and the higher the site index. By holding the age variable constant in site indexing, stands and sites can be compared for their timber productivity. A stand with a site index of 60 was or will be 60 feet tall at age 50, but a similar stand with a site index of 85 will or did achieve 85 feet in height at the same age, and can be said to be growing on a much better site. The site index is therefore an important factor in selecting the stands and sites that have the best potential for timber, sugarbush, or fuelwood production.

The index for a given site will vary among the species that might grow there because each species has particular requirements for nutrients, water, and soil depth. A poor site for one species might be adequate for another. For example, droughty soils with a red oak site index of 45 may have a site index 60 for white pine, which is better able to grow under conditions of low soil moisture. In New England, a site index of 45 or less is poor, 55 to 65 is average, and 80 is excellent.

Recommended Treatments

The basic components of a stand recommendation are an explanation of the purpose of treatments, a summary of treatments recommended for the duration of the plan, and an indication of the priority of the different treatments for the stand.

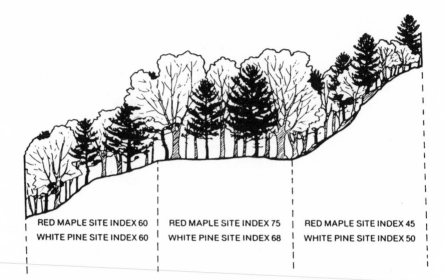

| RED MAPLE SITE INDEX 60 | RED MAPLE SITE INDEX 75 | RED MAPLE SITE INDEX 45 |
| WHITE PINE SITE INDEX 60 | WHITE PINE SITE INDEX 68 | WHITE PINE SITE INDEX 50 |

Fig. 36. Site index varies not only between forest sites, but also between species on the same site. *Left:* site indices for red maple and white pine are equal on the medium site, meaning that when fifty years old the white pine and red maple will be about the same height, on that site. *Center:* on a good site, however, red maple has a higher site index and will be taller than the white pine at fifty years. *Right:* the reverse is true on the poor site to which white pine is better adapted.

Stand Treatment Recommendation

Stand 101

This stand should be regenerated to white pine and mixed hardwoods, with the red maple component reduced. The existing overstory is mature, dominated by trees of poor sawlog quality, and of a single height class. A regeneration harvest designed to create three distinct vertical layers of vegetation is desirable for the improvement of songbird habitat. The missing layers are of vegetation 2 feet or less in height and from 3 to 25 feet tall.

The best way to regenerate this stand while creating the minimum visual impact is to use the shelterwood method. In the first cut, 40 percent of the crown cover of the overstory will be removed by harvesting the hemlock and the least vigorous pines, leaving residual trees at uniform spacing. Approximately 50 percent of the basal area of pole-sized trees in the lower story will be cut for fuelwood. This will add even light to the forest floor that should stimulate the germination of seed already present on the forest floor, and encourage the establishment of a low herbaceous layer. The first cut will yield approximately 3 cords of fuelwood and 2,000 board feet of sawlogs per acre. Assuming current average stumpage prices of $8.00 per cord and $50.00 per thousand board feet for this material, this cut would produce about $124 per acre, or $1,860 for the entire stand.

If an insufficient amount of regeneration is established after the first cut, the second cut should be made in the summer or fall of a high-yield seed year, and should remove approximately one-half of the crown cover of the remaining overstory. Care should be taken to disturb the leaf litter on the forest floor during the logging operation, thus exposing a mineral soil seedbed for the new seed.

After each harvest, all woods roads should be seeded to herbaceous species and maintained at a width that allows sunlight to reach the roadbed. This will ensure the maintenance of the needed lower layer of vegetation. Investment in the control of hardwood sprouts may be necessary to allow successful establishment of the desired quantity of white pine seedlings. Where possible, black cherry should be favored over red maple in order to improve songbird habitat.

The final cut to remove the remainder of the overstory should be made when the white pine regeneration is 12 to 16 feet tall.

Approximately 5 dead or dying trees per acre, which are greater than 8 inches in diameter, should be left for woodpeckers.

Logging contracts governing any operations in this stand should require that all slash be lopped to lie within 2 feet of the ground. This will minimize the visual impact of any operations. If the owner has the time, piling slash will improve the quality of the wildlife habitat in the area.

This treatment is of high priority.

Stand Treatment Recommendation
Stand 102

The stand requires a fuelwood thinning to reduce the density of the trees. The thinning should not occur until gypsy moth egg numbers are low since trees already weakened by annual defoliation may be further stressed by the sudden change in stand conditions that follows a cut. A crown thinning will increase the growth on those trees which are of the most desirable species and form. Only trees competing in the crown should be removed; the small, suppressed trees can remain to shade the site and minimize unwanted regeneration. In this stand the most desirable species is red oak. A fuelwood thinning will remove trees, especially red maple, which are competing with crop trees. Following thinning, approximately 70 square feet of basal area (260 trees per acre) will remain. Trees in the main canopy should be left at approximately 12- to 15-foot spacing. The initial thinning will yield approximately 6 cords per acre. Assuming an average stumpage price of $10 per cord, this treatment would produce $60 per acre, or $1,320 for the entire stand. Approximately 5 snag trees per acre should be left for wildlife. A 25-foot strip of uncut trees should be left adjacent to the stream. Access roads will need to be planned at the time of harvest. This treatment is of medium priority. In 8 to 12 years another thinning will be needed.

The timetable for woodswork is usually a compromise between silvicultural considerations and the needs of the owner, whose financial circumstances may influence the timing of income and expenses. For an owner who intends to do his or her own woodswork, recommendations should be written in understandable terms and in sufficient detail, and should incor-

---

<u>WOODSWORK RECORD</u>

Stand No._____   Forest Type_____

Month(s)/Year:_____

Type of Work:_____

Labor hours:_____

Acres treated:_____

Work done by:_____

Sawtimber removed:_____board feet of_____(species)

_____board feet of_____(species)

_____board feet of_____(species)

_____board feet of_____(species)

_____board feet of_____(species)

_____board feet of_____(species)

Fuelwood removed:_____cords of _____(species)

Pulpwood removed:_____cords of _____(species)

Other products:_____

_____

Residual stand density:_____

☐ Lump-sum sale prices: ☐ Unit prices

    Sawtimber:_____   Species   Price/MBF or cord

    Fuelwood:_____

    Pulpwood:_____

Other:_____:_____

_____:_____

_____:_____

Notes:

Fig. 37. A sample form for keeping a record of woodswork.

porate a realistic time frame. A thinning carried out by an owner will probably take more time than a thinning done by a full-time commercial operator.

Updating Procedures

Management plans provide recommendations for a period of time of usually not less than five years or more than fifteen years. Regardless of the period covered by the plan, however, it should be updated at approximately five-year intervals. State guidelines for management planning under current-use programs sometimes dictate or suggest the frequency at which plans must be rewritten. Updating a plan involves making notations of the changes in each stand's condition, and reconsidering treatments and their timing. When stress from insects and disease is significant, cutting should be postponed. Recording changes is important for evaluating a stand's response to treatment and for monitoring new conditions, such as wind, insect or disease damage. The emergence of new markets may call for adjustments in the treatment of a stand. Keeping a record of woodswork as it is done is useful when it comes time to amend a management plan. A sample woodswork record sheet is illustrated in figure 37.

Many foresters charge a minimum management planning fee for tracts up to a certain size (often 50 or 100 acres), and charge an additional fee for each acre above the minimum. Others charge according to a fee schedule based on the size of the woodland, charging a flat fee for parcels up to 100 acres, another for 100 to 200 acres, and so forth. A few charge by the hour for management planning. If a complete inventory is desirable, its cost should be included in the fee for a management plan (although it is possible to commission a forester to do just an inventory). The cost of map preparation is also included in the fee if the forester has been provided with precise boundary information.

A good management plan contains a record of the purposes and goals of managing woodland, a detailed description of the property, and directions for developing its potential to produce income and enjoyment. Attempting forest management without a plan is akin, in most cases, to taking a long trip through unfamiliar territory without a road map: either the destination may never be reached, or doing so is likely to take much longer than necessary. In either case, the traveler would agree that the cost of a map would have been money well spent.

5 Woodland Management Techniques

Although the objectives of forest management may vary greatly from one ownership to another, the choice of techniques to realize them is surprisingly limited. This chapter describes the techniques most likely to be called for in management plans for private woodland in New England. The discussion is intended to be thorough; it is unlikely that any one management plan would call for all of the treatments described here.

For timber, maple products, fuelwood, pulpwood, or wildlife habitat, a technique is used either to encourage a new stand or to modify the spacing, species, overall quality, ages, or sizes of trees in an existing stand. Techniques used for new stands can be called *reproduction* or *regeneration methods*; those used for stand improvement are *intermediate treatments*.

Most of the techniques employed for either purpose work by manipulating the competition that occurs among the trees on a site. Trees compete with each other for growing space, that is, for the limited supply of light, moisture, and nutrients on a site. When a stand is young and is made up of many thousands of trees, each tree has a small crown and root system and takes up a tiny fraction of the acre and the growing space in which it stands. As the stand grows, the crowns expand, root systems begin to overlap, and competition steadily intensifies. The more successful trees develop larger crowns than their competitors. A tree with a large crown has a greater photosynthetic area with which to manufacture the food necessary for growth.

The process of competition in an even-aged stand leads gradually to distinguishable *crown classes*, as figure 38 illustrates. Trees in the "dominant" crown class stand above the general level of the canopy, receiving light from both the top and sides. "Codominant" trees form the upper level of the canopy; most of their light comes from above, with a minor portion from the sides. Trees in the "intermediate" crown class have noticeably smaller crowns and have visibly suffered from competition with dominants

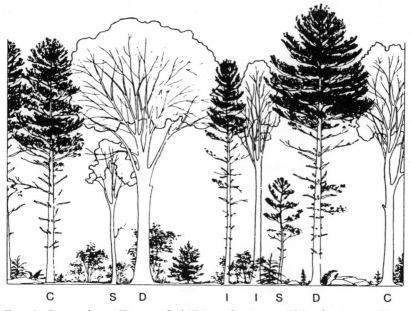

C   S   D   I   I   S   D   C

Fig. 38. Crown classes. Trees marked "D" are dominants, "C" codominants, "I" intermediates, and "S" suppressed, even though all of the trees are in the same age class. The vegetation below the suppressed trees forms the understory.

and codominants; all of their light comes from above. "Suppressed" trees have very poorly developed crowns, receive no significant amount of light, and die relatively quickly. Suppressed trees have lost the race for light, water, and nutrients. Even if released, they usually will not recover. The processes of competition and differentiation into crown classes continues throughout the life of the stand until the population stabilizes at a few hundred trees per acre. In an unmanaged stand, the choice of trees that prevail and those that die is made by chance and natural selection. Woodland management is largely aimed at assuring that the trees most valuable for a given use gain the competitive advantage, and many techniques work by simply removing the undesirable trees, whose share of the growing space is then inherited by the desirable ones.

Techniques for Sawtimber Production

The goal of sawtimber management is to produce high-quality trees of valuable species that reach merchantable size quickly, so the capital represented by them can be recovered as soon as possible. Implicit in any prescription for treatment of a sawtimber stand are two fundamental decisions: which species to favor in management, and which age class to favor.

The decision concerning which species to grow for timber production is based on the suitability of the various species in a stand to the site, on local insect and disease problems, and on potential monetary value. Table 11 lists the minimum site requirements for the adequate growth of commercially valuable New England species.

Species affected by epidemics and persistent local diseases and insect pests may be risky management prospects. Stands with such problems must be managed very carefully so as not to add to the stress by heavy cutting. White pine management must take into account the white pine weevil, and in some areas, a fungal disease called blister rust. Beech has been eliminated from some stands and goes unmanaged in others because of a widespread problem called "beech bark disease." Oak growing on droughty soils is particularly susceptible to repeated defoliation by the gypsy moth. Spruce/fir stands having a large fir component are prime candidates for spruce budworm damage and those too densely stocked are prone to armillaria, a virulent root fungus disease.

Decisions about relative species value are based largely on current market values. As table 12 shows, these values change over time, sometimes quite suddenly. A few examples are worth mentioning. In the early 1970s, the Japanese demand for sugar maple greatly elevated the stumpage price (the price paid to the landowner for trees on the stump) of sugar maple in New England. By the latter part of the decade, however, this market had largely disappeared and stumpage prices fell. The late 1970s saw an explosive growth in the demand for furniture-grade red oak for export to western Europe. Stumpage prices doubled and tripled. During the 1980s, strong Japanese demand for ash developed. By 1990, a recession in Japan had eliminated most of this demand and stumpage prices plummeted. Although it is not a wholly reliable technique, predicting the future value of various species by looking at past trends may be the best method available.

A decision must also be made as to which age class to favor in a stand. If the trees dominating an existing stand are too sparsely distributed, or the stand is otherwise of low value, silvicultural efforts should aim at regenerating the stand. Such a situation might result from past highgrading or from dominance by a so-called "off-site" species that is incapable of producing high-quality sawtimber because of an unfavorable combination of site characteristics. If the existing stand is of high potential value and good

Table 11 Minimum Site Requirements for Adequate Growth of Selected Forest Trees

Species	Moisture	Nutrients	Soil temp.	Light
Balsam fir	H	M	L	L
White spruce	M	M	L	M
Red spruce	M	M	L	L
Red pine	L	L	M	V
White pine	L	M	M	M
Hemlock	M	M	L	L
Larch	H	L	L	V
White cedar	H	M	L	L
Red maple	L	M	M	M
Sugar maple	H	V	H	L
Yellow birch	H	H	M	M
Paper birch	M	M	L	V
White ash	H	V	H	H
Aspen	M	M	M	V
White oak	L	M	M	M
Red oak	M	M	M	M
Basswood	M	V	H	L

Key:
L = low
M = medium
H = high
V = very high

Source: Adapted from U.S.D.A. Forest Service, *Silvics of North America,* Agriculture Handbook 654 (Washington, D.C.; U.S. Department of Agriculture, 1990).

Table 12 Changes in Timber Values, 1954–1978

Species	Rank in 1954	Rank in 1978	Change in Rank
Birch	1	8	−7
Cherry	2	1	+1
Sugar maple	3	6	−3
Yellow poplar	4	10	−6
Basswood	5	5	0
White oak	6	3	+3
Red oak	7	4	+3
Ash	8	2	+6
Red maple	9	7	+2
Beech	10	11	−1
Hickory	11	9	+2

Adapted from: H. V. Wiant, "Shifting Prices and the Silviculturalist," *The Consultant,* vol. 25, no. 3 (July, 1980), p. 82.
Note: Values are ranked from 1 to 11 with 1 being the most valuable and 11 the least.

Table 13 Summary of Intermediate Treatments

Treatment	When Applied	Purpose	Trees Removed	Trees Retained
Release Cutting				
Cleaning	Seedling/sapling	Regulate species composition; regulate tree quality	Unwanted species; trees of poor form	Crop tree species of good form
Weeding	Seedling/sapling	Regulate species composition	Unwanted species	Crop tree species
Liberation	Seedling/sapling	Release young stand	Older, larger over-topping individuals	Seedling/sapling regeneration
Improvement Cutting	Pole and larger	Improve species composition; improve tree quality	Unwanted species; poorly formed trees; overmature trees; injured trees	Well-formed trees of desirable species capable of improved growth
Thinning				
Low thinning	Pole and larger	Salvage mortality; reduce root competition	Individuals of lower crown classes	Individuals of middle and upper crown classes
Crown thinning	Pole and larger	Increase diameter growth rate of crop trees; salvage mortality	Less desirable, poorly spaced individuals of upper classes	Well-spaced, well-formed individuals of upper crown classes
Pruning	Pole, no larger than 8 inches DBH	Improve lumber quality		

vigor, an appropriate intermediate treatment should be considered. In a stand in which two distinct age classes are present, the potential value of each should be considered. Retention and treatment of the overstory may damage the value of the understory. Conversely, concentration on the younger stand might mean forfeiting the potential value of the larger trees if they must be cut prematurely to release the saplings. In some cases cultivation of both age classes will be possible, especially where the lower story is composed of species able to tolerate shade (a characteristic discussed below).

Intermediate Treatments

Intermediate treatments used in the production of high-quality sawlogs include *release cutting, improvement cutting, thinning,* and *pruning.* Brief descriptions of these techniques are given in table 13.

Release cutting.

Release cutting is applied early in the life of a stand, before it passes the sapling stage. Technically there are three kinds of release cutting, but the distinction between the first two—*cleaning* and *weeding*—is narrow. In a cleaning, trees of poorer form and less valuable species that are the same age as the desirable trees are removed from the stand. A weeding is a similar but more specialized treatment in which all trees except those of the favored species are removed. A cleaning seeks to control both the form and the species composition of the young stand; a weeding exerts strict control over the species composition. Both treatments affect the competition within the stand to favor desirable trees. To simplify this discussion, weeding will be included in the more general and more frequently applied technique of cleaning. A stand in which a cleaning would be approprite is shown in figure 39.

The first step in a cleaning is the identification of high-quality trees—those individuals to be favored. Species, form, and spacing are the primary criteria used in the selection process. If multiple sprout stems of desired species are present, a cleaning should eliminate all but the straightest, best-formed members of the clump. Following the selection of crop trees, a rule of thumb is applied to designate the trees to be removed depending on site and stand conditions. For example, the trees to be removed might be all of those within four feet of a crop-tree stem. Cleanings are often made where fast-growing species have overtopped or threaten to overtop a more valuable but slower-growing species. The cleaning allows the desired

Fig. 39. A stand in which a cleaning would be an appropriate prescription in order to produce timber.

species to maintain a competitive position in the stand by giving the crown some room to expand.

However, it is also important that the stand be kept fairly dense at this sapling stage in order to encourage tree form desirable for sawtimber production. A dense stand forces trees to grow straight and tall, and encourages self-pruning when competition from neighboring trees results in the death of lower side branches. Such pruning is desirable because lower branches result in knots in lumber. A dense stand helps to support the trees and affords some protection from snow and ice damage. A final reason for keeping a sapling stand dense is to offset anticipated mortality. Some stems can be expected to succumb to various natural forces as the stand develops. The sapling stage is too early a time to make a final and irreversible selection of crop trees; enough trees should be left after the final cleaning to provide for natural losses, at least one thinning, and a sufficient final harvest.

Cleanings are usually accomplished by felling the unwanted trees and leaving them on the ground. The wood from the trees killed is of such small diameter that economic removal is not generally feasible. Also, the operation of equipment to remove the wood in a dense sapling stand may damage enough of the residual trees to negate the value of the cleaning. Landowners who do their own woodswork may find it feasible to remove some of the wood produced in a cleaning if a stand is close to a house, and especially if they use small fuelwood.

When felled trees are left on the ground, the operation is termed "precommercial"; it represents a cost, rather than a revenue. In some areas cost-sharing money from the federal government may be available to cover part of the expense of cleaning. Because most cleaning operations are precommercial and necessitate spending money or a good deal of time in a sapling stand, the need for cleaning should be clearly established before the operation is undertaken. Many sapling stands are of sufficiently good form and composition to develop satisfactorily without a cleaning.

The third type of release cutting is called *liberation cutting*. This treatment differs from cleaning in that the trees removed overtop the stand and are older than those in the stand that are being favored (fig. 40). Their removal makes available to the younger trees a considerable amount of light, moisture, and nutrients.

Liberation cuttings are most commonly needed when large, poor-quality trees or inferior species overtop and dominate a sapling stand of greater potential. In New England, highgrading or a major storm may have left an overstory in poor condition, or natural succession may have placed a poorer species above a more valuable one. Liberation cuttings are often needed to remove *wolf trees*—a legacy of New England's forest history. These are frequently trees of poor form or less valuable species which were

Fig. 40. A stand of white pine overtopped by white birch and aspen. In order to manage the white pine for sawtimber production, the appropriate prescription would be a liberation cutting to release the white pine by removing the white birch and aspen.

left behind during past highgrading operations and are now of considerable size. They consume a large share of the growth potential on a site and are therefore "predators" of younger trees beneath them (see fig. 41).

When a liberation cutting is prescribed, there are three potential methods of doing the work. The first is to cut, remove, and sell the trees killed. This is feasible if the trees have any value and if they can be removed without undue damage to the young stand beneath. The second method is to fell the trees and leave them on the ground. This procedure causes some damage to the young stand, but not nearly as much as trying to extract the wood. With the third method, the overtopping trees are girdled in place. Girdling involves cutting away a ring of live tissue around the tree to block the flow of nutrients between the leaves and the roots.

Girdling can be accomplished by a chain saw, an axe, or a hatchet (figs. 42, 43). Chemical poisons can be used instead of, or in combination with, a complete girdle. However, because of controversy over the safety of chemicals, many landowners and foresters prefer a completely mechanical approach to girdling. The advantage of using chemicals is that they prevent hardwoods from sprouting following treatment. On the other hand, "flashback" may result in the death of trees that were to be retained; flashback occurs when chemicals are transferred from one tree to another across grafted roots. *The issues surrounding the use of chemicals are complex and should be carefully evaluated.* Their use on private land in New England does not seem to be widespread at this time.

Although some landowners find the sight of girdled trees distasteful, this method of liberating a promising understory offers three advantages. As the trees die and rot, they often break up gradually, thereby minimizing damage to the stand beneath. Second, girdled trees become dead standing trees, or snags, which are important to many species of woodland wildlife for food and cover. Third, where an overstory is unmerchantable, girdling

Fig. 41. A wolf tree. Such remnants of previous stands present a hard choice between maintaining them for their beauty and wildlife value, or removing them to release the young stand below.

Fig. 42, 43. Girdling done with a chain saw (fig. 42) and an axe (fig. 43). Girdling cuts must ring the tree completely, severing the bark and up to an inch of wood, and be continuous. A double cut with a chain saw is more effective than a single ring.

is much more economical than felling or removing it. Federal cost-sharing money may be available to cover some of the cost of girdling an unmerchantable overstory.

Improvement cutting.

An improvement cutting is made in the overstory of a stand that is of pole size or larger which has received no previous management. Its purposes are to regulate species composition and to improve overall tree quality and vigor by removing undesirable trees. The trees removed are of poor form, vigor, quality, or species; usually, they will be as large or larger than the trees retained.

Because of past patterns of land use, improvement cuttings are very important in sawtimber management in New England. Many of the stands in the region are of pole size and larger, having originated after heavy cutting or natural disturbances during the first half of the century. Most of these stands received no early management.

On typical New England properties, it may take from ten to fifteen years to put a previously unmanaged stand into good growing condition. For many reasons, it is rarely good practice to cut so heavily during an intermediate treatment that a large flush of sunlight is introduced into a stand. Especially in a previously unmanaged stand, sudden exposure may result in the formation of undesirable lower branches, called *epicormics*, and in the establishment of unwanted understory species. Because epicormics form knots, they can degrade the lumber in a tree. Oaks and sugar maple are particularly prone to epicormic branching.

It is important that the trees to be left following an improvement cut will be capable of good growth, and that they have potential as good sawtimber trees. All too often improvement cuts are attempted where the residual stand is too old and of such poor quality that it cannot be improved. Where this is the case, consideration should be given to establishing a new stand rather than attempting to work with the existing one.

Most improvement cuttings are commercial: they produce some revenue from fuelwood, pulpwood, sawlogs, or a combination of these. In addition to improving the quality, condition, and composition of a stand, improvement cuts allow the landowner to recover some of the value of the excess wood that has built up during the earlier life of the stand. Occasionally, an improvement cut is a precommercial operation in which the unwanted trees are girdled, but if a stand is of such poor quality that a commercial operation cannot be supported, it is doubtful that the improvement cut will benefit the residual stand. *Regeneration*—removing and replacing the stand—is probably a more viable option in such a situation.

Thinning.

Like improvement cuts, thinnings are conducted in stands that are of pole size and larger. The term "thinning" is often used loosely to describe any intermediate cut. It is used here to refer to a cut undertaken for two specific purposes: the primary purpose is to increase the growth rate of the diameter of the remaining trees; the secondary purpose is to capture those trees which, if left in the stand, would die from suppression. A thinning of this kind differs from an improvement cut in that the main considerations in tree selection are not quality and species composition, but tree spacing for optimum growth. This fine distinction is usually blurred in practice in New England, however, because thinnings and improvement cuts are usually applied simultaneously.

The primary purpose of a thinning is to transfer the growth potential of the trees removed to the remaining crop trees, thereby increasing their diameter growth rate. Crop trees then reach their targeted harvest diameter in a shorter period of time. An auxiliary financial benefit is the generation of income before the final stand harvest, because the trees removed may be usable for fuel or pulp, or as small sawtimber.

The secondary purpose of thinning—to recover the wood volume in the trees that would otherwise be suppressed—effectively increases the productivity of a stand. This does not mean that thinning increases the total amount of wood grown on a site; the increase is in the total amount that is recovered for use. It is worth repeating that the potential for yearly fiber production on a given acre is essentially fixed, and is basically a function of site characteristics such as soil depth, fertility, and moisture. The promise of timber management is not to increase growth, but to increase the amount of growth that can be made into valuable products.

A series of light thinnings can increase the windfirmness of trees, thereby lowering the risk of losing a whole stand in a major storm. The development of a full, balanced crown and a strong root system increases the ability of a tree to withstand major winds. There are situations, however, in which thinning will reduce the windfirmness of a stand. On wet soils with high water tables, trees tend to be shallow-rooted and receive less support in high winds than trees on deeper soils. Thinning removes trees that are instrumental in reducing sway in high winds, and in very dense stands where trees are abnormally tall for their diameter, a thinning may leave them vulnerable to whipping action in high winds. At least in the first thinning of a stand, the trees along borders with open areas should be retained as a wind buffer for the remainder of the stand. In general, the effect of a thinning on windfirmness depends on a variety of factors, including the proportion of the stand removed, the species of the residuals, soil characteristics, aspect, the intensity of prevailing winds, tree form, and past treatment of the stand. The feasibility and intensity of thinning has to be evaluated on a stand-by-stand basis.

Thinnings are frequently classified as commercial or noncommercial on the basis of whether the trees they eliminate can be removed profitably. The term "precommercial" is sometimes used to signify a noncommercial thinning or another treatment in which the products removed are too small to be saleable. Thinnings are usually commercial where and when there is fuelwood demand. Noncommercial operations are almost always accomplished with girdling, and are necessary when there is no market for the material to be thinned from the stand; when stands are so remote that costs to get the wood out of the forest exceed value; and when tree diameters are too small or quality is too low to permit economic removal. The fuelwood market and the development of new markets for chipped wood are reducing the applicability of girdling in thinning.

Commercial thinnings are often categorized according to the products they remove: fuelwood, pulpwood, or sawlog thinnings. A less commonly heard categorization rests on the relative height of the trees removed—low thinnings and crown thinnings. A low thinning removes trees whose crowns are below the general level of the canopy in order to reduce root competition with the overstory crop trees and to capture the wood volume of those trees likely to die from suppression. A crown thinning removes noncrop trees whose tops are in the canopy and are interfering with the crown spread of crop trees (figs. 44, 45). Crown thinnings are considered more effective than low thinnings. Since an increase in diameter growth is most likely when the crowns of crop trees are given room to expand, low thinnings do not substantially modify competition in the canopy. In practice some low thinning to salvage volume that would be lost is usually done during a crown thinning.

Thinning is the most common prescription for sawtimber production in management plans for New England woodland. It is prescribed repeatedly

Fig. 44, 45. Crowns of white pine in an unmanaged stand (fig. 44) and in a thinned stand (fig. 45).

over the life of a stand managed from youth. Thinnings should be applied whenever overcrowding causes decelerating rates of diameter growth (fig. 46). A program of thinnings carried out during the pole and sawlog phases of a stand serves to maintain a fairly even growth rate. Many New England species, if suppressed for many years in an overcrowded stand, will not increase their growth rate in response to thinning. In other words, maintaining a stand's growth rate is similar to maintaining the forward motion of a car up a snow-covered hill: as long as momentum is maintained, progress is smooth, but if the car is forced to stop, it may be difficult to resume movement, except at a very slow speed. A few very shade-tolerant species such as sugar maple, spruce, and especially hemlock are the exception since they often respond to release at advanced ages.

Each thinning involves further refinement in the selection of crop trees, removes trees with less potential that impede crop tree growth, and provides favored trees with sufficient growing space. In no thinning are so many trees removed as to increase the risk of blowdown or encourage lower branches to form. In previously unmanaged hardwood stands, heavy thinnings encourage the development of epicormic branches.

Thinning should encourage ideal tree form, which represents a compromise between the need for a well-developed crown to fuel rapid growth and the need for a straight bole free of branches to provide good lumber quality. The expected frequency of thinning affects the number of trees removed at any one time. Ideally, thinnings should be light and frequent, creating enough space for crown expansion for a few years at a time. Such a conservative approach discourages unwanted shrubs and regeneration because light levels on the ground remain low. When the crowns close together, another thinning is required. Unfortunately, from a silvicultural standpoint the ideal thinning cannot always be made economically. As a rule, to be economically viable for a logger a thinning must remove at least 5 cords, or 2,000 board feet, or an equivalent combination per acre.

Pruning.

Pruning is one intermediate treatment that does not involve eliminating trees. It removes live or dead branches for the purpose of preventing knots in lumber (fig. 47). In New England it is applied most commonly in softwood stands, especially those of white pine. Hemlock, spruce, and other species grown to produce two-by-fours and similar dimension lumber used in construction are seldom pruned, since the value added by pruning does not justify the cost. Hardwoods are seldom pruned for timber production in New England because in forest conditions they are less likely to have persistent lower branches, and their relatively slow growth makes the term of pruning investment too lengthy.

In timber management, pruning is most often undertaken in pole-size stands at a time when crop trees can be fairly well identified. Branches are removed with a special pruning saw, theoretically up to 11, 17, or 25 feet

Fig. 46. A simplified drawing of an increment core displaying the effects of competition and thinning. As the crowns and roots of the trees in the stand begin to crowd each other, competition intensifies and growth slows down at C. Ideally, a thinning would be applied at point T, reducing the crown and root competition and allowing resumption of a rapid growth rate.

Fig. 47. The effects of pruning. *Left:* knots in lumber from trees that were pruned when young are small and are limited to an interior portion of the sawlog. Since pruning, the tree on the left has added several inches of clear (knot-free) wood. *Right:* unpruned branches often create large knots that extend outward from the center to the exterior of the sawlog, resulting in little or no clear lumber. Knots are structural defects and sometimes visually undesirable.

above the ground in order to assure high-quality logs of 10, 16, or 24 feet in length. However, the wisdom of pruning higher than 17 feet is questionable; pruning farther up a tree involves the use of ladders, which greatly increases the investment in time in an already labor-intensive process.

No more than about 100 trees per acre should be pruned: only this number will survive long enough in a managed stand to add enough growth of clear, knot-free wood to justify the pruning investment. Trees selected should be from 4 to 6 inches in diameter, and never larger than 8 inches. They should be straight and defect-free, with healthy, well-proportioned crowns. Pruning should follow, rather than precede, thinning, since some trees will usually be damaged in thinning operations or blow down in severe storms that occur shortly after thinning. Because labor costs are so high, pruning has become less common than it used to be. The potential return from the practice is still hotly debated among foresters and sawmillers.

Regeneration Methods

Like all living organisms, a stand of trees reaches an age at which its growth rate levels off. When a stand's annual growth begins to drop below its lifetime average of annual growth, the stand is considered to be biologically mature (fig. 48).

For a period of time after biological maturity occurs, the slowing of the growth rate may be offset by the fact that what growth does occur is of greater value per board foot (or other unit of measure). Growth added to a larger tree is often more valuable than an equivalent amount of growth added to a smaller tree because the wood is more likely to be knot-free, a large tree can yield wide boards—which are more valuable per unit mea-

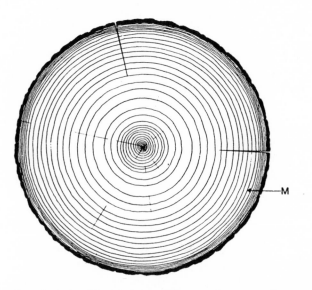

Fig. 48. A simplified drawing of the growth pattern of a tree reaching maturity. A tree growing in ideal conditions, with no competition from other trees, would display a diameter growth pattern that included an establishment phase of relatively slow growth, rapid midlife growth, and declining growth as the tree reaches old age. A tree is considered biologically mature when, despite adequate growing space, the diameter growth added each year is consistently less than the average growth to date. In the drawing, growth averages about .15 inches annually. At about point M, annual growth falls below that average as the tree approaches maturity. Note that because of competition and other environmental conditions, forest trees rarely display such a regular growth pattern. Also, note that a mature tree would be much older; for simplicity, only a few growth rings are shown.

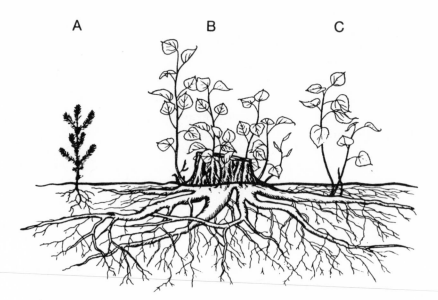

Fig. 49. A seedling (A), stump sprouts (B), and root suckers (C).

sure than narrow boards—and logs of larger diameter can be sawn with much greater efficiency (see chap. 7).

A stand is "financially" mature when the dollar value of its current annual growth drops below the average annual value of its growth to date. If financial considerations are paramount, when a stand reaches financial maturity it is time to regenerate the stand with a harvest, and turn the focus of management to its successor.

In any regeneration method, the next generation of trees might arise as seedlings, stump sprouts, or root suckers (fig. 49). Where sawtimber production is the management objective, it is usually desirable for the new forest to begin from seed because fungi, which decay stumps and roots, can spread into suckers and sprouts, and because trees that originate as sprouts are often multistemmed and of poor form (fig. 50). Furthermore, seedling regeneration will have greater genetic diversity since the seedlings will be the offspring of two parents, while sprouts and suckers will have exactly the same genetic makeup as the single parent tree. Maintenance of genetic diversity promotes a variety of adaptations to stresses such as disease infestation.

However, sprouts and root suckers grow faster than seedlings because they have the benefit of a large, already developed root system. Almost all New England hardwoods sprout following cutting, as do pitch pine and (perhaps) northern white cedar. Aspen and beech are the species most likely to produce root suckers. Stump sprouting ability declines as the size of the parent tree increases, and varies among species.

Success in promoting the germination and growth of seedlings of desired species depends on a number of factors, including chance. Of primary importance are the availability of seed and the proper environmental conditions for germination. Table 14 lists the soil, light, and moisture requirements for seedlings of major tree species. It can also serve as a guide to the conditions to be maintained or created when an overstory is removed.

Regeneration methods work largely by manipulating light levels. Attention to the shade tolerance, or the ability of a species to withstand shade, is particularly critical. *Shade-tolerant* (*tolerant*) species are those that can survive in shade; those that require full sunlight in order to survive are known as *intolerant*. Shade tolerance ratings for major New England tree

Fig. 50. A stand of trees that originated as stump sprouts.

Table 14 Optimum Conditions for Regeneration of Major Tree Species

Species	Preferred Seedbed(s)	Light Level	Soil Moisture Level
White pine	Mineral soil[a]; moss; sod	Partial shade	Dry to moist
Red pine	Mineral soil	Partial shade to full light	Dry to moist
Pitch pine	Mineral soil	Full light	Dry to wet
Hemlock	Mineral soil; rotted wood; litter[b]; moss	Partial to full shade	Moist to wet
Balsam fir	Duff[c]; mineral soil; rotted wood	Partial shade	Moist
Red spruce	Mineral soil; rotted wood; duff	Partial shade	Moist
White spruce	Rotted wood; mineral soil	Partial shade	Moist
Atlantic white cedar	Rotted wood; peat; moist mineral soil	Partial shade to full light	Moist to moderately wet
Northern white cedar	Rotted wood; organic soils[d]	Partial to full shade	Moist to wet
Eastern red cedar	Sod; mineral soil	Full light to light shade	Dry to moist
Larch	Mineral soil; organic soils	Full light	Moist to wet
Red maple	Mineral soil	Not critical	Dry to wet
Sugar maple	Mineral soil; litter	Partial shade	Moist
Silver maple	Mineral soil	Full light	Moist to wet
Paper birch	Mineral soil; rotted wood	Full light	Moist
Black birch	Mineral soil; rotted wood; duff	Partial shade	Moist
Yellow birch	Mineral soil; duff	Partial shade	Moist
Beech	Mineral soil; litter	Partial to full shade	Moist
Ash	Mineral soil	Full light to light shade	Moist to moderately wet
Basswood	Mineral soil	Partial shade	Moist
Black cherry	Litter; duff; mineral soil	Partial shade	Moist
Butternut	Litter	Full light to light shade	Moist
Red oak	Litter	Partial shade to full light	Moist

Species	Seedbed[a][b][c]	Light	Moisture[d]
Black oak	Mineral soil	Partial shade to full light	Dry to moist
Scarlet oak	Litter	Partial shade to full light	Dry to moist
Chestnut oak	Litter	Partial shade to full light	Dry to moist
White oak	Litter	Partial shade to full light	Dry to moist
Shagbark hickory	Duff; litter	Partial shade	Moderately moist
Other hickories	Duff; litter	Full light	Dry to moist
Yellow poplar	Mineral soil	Light shade	Moist
Aspen; poplars	Mineral soil	Full light	Moist
Elm	Litter; moss; rotted wood	Partial shade	Moist

Source: Adapted from U.S.D.A. Forest Service, Seeds of Woody Plants of the United States. U.S.D.A. Forest Service Ag. Handbook 450 (Washington, D.C., 1974).

[a] Mineral soil seedbeds can be created by scarifying, or disturbing the soil during or after commercial harvesting operations.
[b] Litter is the leaf cover on the forest floor in an undisturbed stand.
[c] Duff is the organic, partly decayed soil layer just below the litter.
[d] Organic soils are found in poorly drained areas; they suffer from low oxygen content.

Table 15 Shade Tolerance of Selected Forest Trees

Conifers	Hardwoods
Very Tolerant	
Eastern hemlock	Eastern hophornbeam
Balsam fir	American hornbeam
Atlantic white cedar	American beech
	Sugar maple
Tolerant	
Red spruce	Silver maple
Black spruce	Basswood
White spruce	
Northern white cedar	
Intermediate	
Eastern white pine	Yellow birch
	Black birch
	American chestnut
	American elm
	Red maple
	Ashes
	Oaks
Intolerant	
Eastern red cedar	Black walnut
Red pine	Butternut
Pitch pine	Hickories
	Paper birch
	Yellow poplar
Very Intolerant	
Larch	Willows
Jack pine	Aspens
	Cottonwood
	Grey birch
	Black locust

species are given in table 15. The table shows that there are species which can withstand some degree of partial shade. It is important to understand that no species can grow without some sunlight; tolerant species can withstand heavy shade only for some portion of their lives. The strategy of the tolerant species for survival seems to be to bide their time in the shade of the overstory until the overstory deteriorates or is leveled; then the tolerant species inherit the site. The strategy of the intolerants is to rapidly colonize open, sunny areas and to grow quickly before other species can take hold. Intolerant species are also called "pioneer" species. The importance of tolerance in natural succession can be seen in the forest history narrated in chapter 1. Because regeneration methods seek to imitate natural processes, the concept of tolerance is pivotally important in their application.

To regenerate many New England species, the leaf litter on the forest floor and the soil surface must be deliberately disturbed (*scarified*) because seeds have difficulty sending roots through a compact layer of leaves or needles, and must be in contact with the mineral soil that is beneath the leaves and decomposing plant matter. A certain amount of scarification occurs unintentionally during harvesting operations, especially when trees are skidded out of the woods. Scarification can be increased by logging on snow-free, unfrozen soil, and by dragging an implement or a log behind a piece of harvesting equipment. Scarification should result in widespread, superficial mixing of the litter, duff (partly decayed organic matter), and mineral soil, without deeply disturbing the forest floor.

In theory, there are four techniques for reproducing or regenerating a stand of trees; these are largely distinguishable by the combinations of light, temperature, and moisture that they aim to create on the forest floor. Basically, the methods are clearcutting, seed-tree cutting, shelterwood, and selection cutting. The clearcutting, seed-tree, and shelterwood methods of regeneration establish new stands that are even-aged; they are applied within a relatively concentrated time period so the trees in the new stand will be of approximately the same age, and will be cut at the end of a rotation. A rotation is the total number of years a stand will take to reach maturity and final harvest.

Selection cutting is the only regeneration method intended to create and maintain an uneven-aged stand—a stand in which at least three different age classes are intermingled. Table 16 summarizes the appropriateness of each regeneration method for major New England timber types.

The clearcutting method: complete overstory removal.

Complete overstory removal is accomplished by clearcutting: all trees in a stand that are greater than 2 inches in diameter are removed at once (fig. 51). There are two distinct situations in which clearcutting is appropriate. In the first, advanced regeneration is established. An understory composed of varying proportions of herbaceous plants, shrubs, and seedlings of toler-

Table 16 Suitability of Harvesting Methods for Natural Regeneration of Major Timber Types

Forest Type	Harvesting Method			
	Clearcut	Seed Tree	Shelterwood	Selection
Spruce/fir	Good, if advanced regeneration is established	Poor	Good, especially if no advanced regeneration is established, but not on shallow or wet soils	Good, especially for spruce, but not on shallow or wet soils
Northern hardwoods	Good (1) when regeneration from sprouting is desired; or (2) when advanced regeneration is established; or (3) for less tolerant associated species, such as paper birch, red maple, black cherry, white ash	Poor, except for paper birch	Good, unproven for paper birch	Good for sugar maple and beech
Central hardwoods	Good (1) when regeneration from sprouting is desired; or (2) when advanced regeneration is established	Poor	Good, especially if no advanced regeneration is established	Poor
White pine	Good, if advanced regeneration is present	Poor	Good	Poor
Hemlock	Good, if advanced regeneration is present	Poor	Good	Good

Sources: Frank and Bjorkbom, *A Silvicultural Guide for Spruce-Fir in the Northeast* (Upper Darby, Pa.: U.S.D.A. Forest Service, 1973); Leak, Solomon, and DeBald, *Silvicultural Guide for Northern Hardwood Types in the Northeast*, revised (Broomall, Pa.: U.S.D.A. Forest Service, 1987); D. E. Hibbs and W. R. Bentley, *A Management Guide for Oak in New England* (Storrs, Ct.: Cooperative Extension Service, 1983); K. F. Lancaster and W. B. Leak, *A Silvicultural Guide for White Pine in the Northeast*, General Technical Report NE–41 (Broomall, Pa.: U.S.D.A. Forest Service, 1978); Society of American Foresters, *Choices in Silviculture for American Forests* (Washington, D.C.: Society of American Foresters, 1981); U.S. Forest Service, *Silvicultural Systems for the Major Forest Types of the United States*, Agricultural Handbook no. 445 (Washington, D.C.: U.S.D.A. Forest Service, 1973)

Fig. 51. A clearcut. In this clearcut, all trees have been removed, which is ideal. It is acceptable to leave trees less than 2 inches DBH, however. This photograph was taken in May, four months after logging. The same area is shown five months later in fig. 57.

ant and semitolerant trees often develops in the understory of a stand. If seedlings of tree species become well established in large enough numbers, such advanced regeneration will determine or heavily influence the composition of the next stand. If this understory is vigorous and contains desirable species, clearcutting can be used to release it, thus forming a new stand.

Clearcutting is also the most appropriate method for regenerating intolerant species. In this case, overstory removal can be carried out before the seedlings of the new stand are present—that is, without advanced regeneration. In fact, because of their intolerance, some species will not reproduce until the overstory has been removed. After clearcutting, regeneration depends on nearby seed sources, seed already present in the soil, seed present in the harvested trees, or from stump sprouts or root suckers.

A clearcut for intolerant species must be large enough for the temperature and the amount of light (the microclimate) in the opened area to remain unaffected by the shade of surrounding trees—that is, at least 50

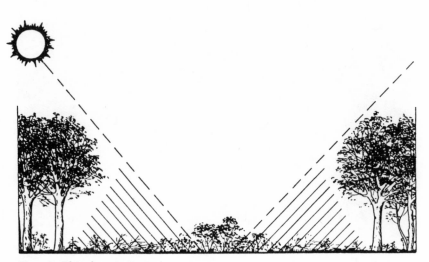

Fig. 52. This clearcut is too small: more than half the area is significantly shaded by adjacent trees.

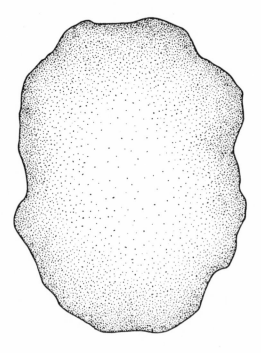

Fig. 53. This clearcut is too large if seed from adjacent stands is the anticipated source of regeneration because seed cannot be adequately distributed throughout the clearcut area, as illustrated by the dot pattern.

percent of the cut area should have the same climate as any larger clearing. If too small, an opening is shaded by adjacent trees and does not encourage the intolerant species (fig. 52); if too large, the center of a clearcut area may be inadequately seeded by adjacent stands if wind-disseminated seed is the expected source of regeneration (fig. 53). For adequate regeneration, the maximum size of a clearcut is not a consideration if planting, stump sprouts, or root suckers are the expected sources of regeneration. The anticipated sources of regeneration to an area should strongly influence the size, shape, and layout of a clearcut for intolerant species.

The use of clearcutting on small, private woodland holdings in New England has been limited for aesthetic reasons, and because the extensive movement of equipment in a clearcut raises the threat of erosion on steep hillsides. Erosion does not result from the cutting of trees, but rather from disturbance of the soil by logging equipment on steep slopes.

To regenerate more tolerant species and forest types, clearcutting should not be undertaken unless and until an acceptable amount of desirable advanced regeneration, from 2 to 4 feet tall, is present. Some species do not survive full exposure as very young seedlings, and require the protection of an overstory until the seedlings are well established. Also, the more tolerant species need a head start to compete with intolerant seedlings that may invade an area once the overstory is removed.

Clearcutting where advanced regeneration is present is similar to both a liberation cutting and the final cut in a shelterwood. In a liberation cutting, however, the overstory is usually sparse or of poor quality or unwanted species, and such a cutting is usually either marginally commercial or noncommercial. For clarity, we here define an overstory removal with established advanced regeneration as a clearcut, if the stand has received no treatments specifically designed to obtain the advanced regeneration.

The seed-tree method: semicomplete overstory removal.

Semicomplete overstory removal by the so-called seed-tree method retains from one to ten trees of good vigor and seed-bearing age on each harvested acre to reseed the area; all other trees are removed (fig. 54). The seed trees are either removed within a few years of the harvest, when seedlings are present, or left on the site until the harvest of the new stand. Crowns of seed trees should be large enough to produce substantial seed crops.

The seed-tree method is suited to the regeneration of intolerant species,

Fig. 54. An area being regenerated by the seed tree method.

although it should not be used where soils are shallow or wet, or in stands of shallow-rooted species, since the seed trees will be susceptible to windthrow. Theoretically, its advantage over clearcutting is that the next generation is not so entirely dependent upon seed already on the ground or on seed that travels into the area from adjacent stands. The seed source represented by the seed trees will ideally be available for at least a few years, until reproduction of the desired species is successfully established.

In New England, use of the seed-tree method is uncommon, since experience has shown that adequate regeneration of intolerant species is usually obtained by clearcutting, which is more efficient, and that the seed-tree method does not provide enough of an overstory for use in regenerating other species. Unless seed tree removal is performed very carefully, a new stand may suffer excessive damage. If the seed trees are of insufficient volume or value to warrant their harvest, they are left in place and not removed until the new stand is mature, in which case they may take up valuable growing space. In considering the seed tree method, its considerable demands must be weighed against what is often in New England only a negligible advantage in control of species composition.

The shelterwood method: gradual overstory removal.

Gradual removal of the overstory is known as the *shelterwood method*. Securing advanced regeneration before the overstory is removed is a cornerstone of this regenerating technique, which is best suited to tree species with a medium to high shade tolerance. The overstory is not completely removed until adequate advanced regeneration of the desired species is established to form the new stand. The shelterwood method involves a sequence of treatments which are applied as the overstory nears maturity (fig. 55). A first removal (the preparatory cut) eliminates from 20 to 40 percent of the canopy. Less vigorous trees and trees of unwanted species are taken, adding light to the forest floor, stimulating seed production by residual trees, eliminating unwanted seed sources, and creating an environment for seed germination.

The remaining trees are then removed in two or more stages, with the trees of best quality retained at each stage. The second cut, called the seed cut, is made in a good seed year; it provides additional light for seedlings which will soon germinate. Summer logging is recommended for a seed cut if scarification is desirable. It eventually may be necessary to plant seedlings or broadcast tree seed if a good seed year fails to occur.

Fig. 55. An area being regenerated by the shelterwood method. The photo shows the area with regeneration well established, just prior to the final cut to remove the residual overstory. *Photo courtesy USDA Forest Service.*

The final overstory removal can be made in one or several cuts, but it should not be made until advanced regeneration of desirable species is present in the understory in sufficient size and quantity. The number and size of seedlings necessary for adequate advanced regeneration varies among species. Seedlings should be uniformly spaced throughout the stand, and there should be many more than are necessary to stock the final stand. Mortality can be expected from logging, competition, and a variety of natural causes. For all species, the final cut should not be made until desirable seedlings are at least 2 to 4 feet tall, since at that stage of development they have strong root systems and a better chance of surviving a logging operation. In the case of white pine, final removal should be delayed until the regenerated trees are from 12 to 16 feet tall to allow for one good log should the white pine weevil, which thrives in full sunlight and attacks the tree's top branch, or "leader," later attack the stand. Winter logging may be optimal once regeneration has been established because snow cover helps to protect it and because hardwoods sprout more vigorously from winter than from summer damage.

In a "two-cut" shelterwood method the preparatory cut is dispensed with and the initial cut is made in a good seed year. This variation is well suited to northern hardwood stands having a large proportion of sugar maple, and to stands of spruce, fir, hemlock, and white pine. The first cut removes from 30 to 60 percent of the crown cover, depending on the species and on the extent of advance regeneration already present. The second, and final, cut is made when regeneration has been successfully established.

Although still unproven for a few species, the shelterwood method is increasingly used to regenerate hardwoods in New England, and is very well suited to the regeneration of white pine and red spruce. Shelterwood has several advantages over the other two even-aged regeneration methods. It provides a gradual rather than an abrupt change in the appearance of a forest being harvested. It provides increments of light to understory seedlings, and yet protects them from sudden exposure. It spreads the income from the final harvest over a longer period, and allows for more control over the composition of the new stand.

The shelterwood system has disadvantages, however. At least one of the cuts must be timed to occur in a good seed year (not always predictable), regardless of the economic climate at the time. It may be difficult to find a logger willing to take only the smaller and least valuable trees in the initial cuts. Unmerchantable small trees that should be removed may have to be girdled or felled in a follow-up operation. Shelterwood requires great care during logging to protect both residual overstory trees and the new stand. On some soils, windthrow in the residual overstory trees could be a hazard. It may be necessary to invest in the control of unwanted hardwood species between cuts in the overstory. The owner must be willing to forestall harvest of the most valuable trees until advanced regeneration is established. Finally, the shelterwood system may require conscientious attention to the task of regeneration over a long time period—possibly up to 20 years, although the period is usually shorter.

The selection method: continuous overstory removal.

In theory, the selection method works by removing the oldest age class in a stand in such a way as to encourage its replacement by new regeneration. Simultaneously, appropriate intermediate treatments are applied to the younger age classes. This technique contrasts with the three methods for regenerating even-aged stands, in which most intermediate treatments are intended to improve only the trees in a single selected age class (usually the overstory). The point of the selection method is to make an ideal apportionment of the growth potential of a site among all age classes, including those in the understory layers, thereby forming an uneven-aged stand in which a few high-quality, mature trees and some younger pulpwood or fuelwood trees are available for harvest from each acre at relatively short intervals. The selection method does not entail rotations; the stand is continuously regenerated, and timber stand improvement is done simultaneously. The periods between selection cuts are called *cutting cycles* instead of rotations. The selection method is more than just a regeneration technique; it is really a complete management system.

The selection method is easiest to apply in existing uneven-aged stands. However, because of New England's history of extensive farmland abandonment and clearcutting, and because true uneven-aged stands rarely occur naturally, there are probably few such stands in the region. They are also difficult to recognize visually, since a small tree is not necessarily a young tree. Practically speaking, then, the selection method is best applied in stands of very tolerant species that have the capacity to respond to release at relatively advanced ages. Such species are sugar maple, beech, hemlock, and spruce/fir. The ability to respond to release means that smaller trees as old as 70 years can be counted on to move into larger size classes with each selection cut. With less tolerant species, use of the selection method in an even-aged stand may be disastrous because it is possible to eventually run out of trees to cut. Smaller trees of the less tolerant species, once suppressed, are unlikely to respond to release and do not regenerate in shade. Oak is a typical species in this regard; other less tolerant species, such as pine and paper birch, react in the same way to a greater or lesser degree.

Of more importance than stand and individual tree age in shaping a stand by the selection method are the distribution of tree diameters in the stand and the number of trees in each size class. Diameter distribution is important because trees have to be maintained in all size classes to ensure the availability of some large trees at each harvest. The silvicultural purpose of each harvest is to reduce the number of trees in each size class to an optimum density for good growth.

In theory, the selection method is appealing in many ways. It provides a relatively stable forest environment (a result of the retention of a continuous forest cover) and thus has aesthetic value and advantages for wildlife. Because trees of a variety of sizes and ages are less likely to be decimated by insects and disease than are stands of uniform maturity, the selection method provides some risk protection. The most attractive potential of the selection method is a sustained yield. If a stand is manipulated so that all size classes are eventually represented, from the current year's reproduction to fully mature sawtimber, then a number of large trees can be harvested every year. (Typically, with the selection method, harvesting is done at longer intervals of 10 to 20 years to ensure harvests large enough to be worthwhile.) Also available for harvest at each cutting is whatever volume had developed in each diameter class in excess of the optimum for good growth.

Like all regeneration methods, the selection method has its drawbacks, but several are more serious than those of other methods. Converting a stand to the proper diameter distribution may take a long time (from 20 to 40 years or more), and there may be a few harvests during which per-acre yields of merchantable timber are too low to yield a profitable sale. Yields may be composed of trees of many sizes, which can make marketing difficult. Some buyers balk at logging jobs that require them to deliver logs to several markets because profitability is reduced. Economic considerations sometimes rule out cultural work in the small-size classes if the trees are not large enough to be saleable. Of all regeneration methods, selection probably results in the most damage to the residual stand (all other factors being equal); very careful logging is required.

The selection method is complex, requiring much experience, sound judgment, singleminded attention, and record keeping. The ability to determine the cutting cycle may require a more thorough knowledge of the factors of species biology, markets, logging systems, growth, and site than is needed for other methods. The cycle is determined by a series of calculations and estimations to arrive at the length of time that the stand, maintained at optimum stocking, will take to grow enough volume to afford a profitable sale. Optimum stand stocking must be predetermined, then apportioned among all size classes (table 17). When marking trees in the selection method, a constant check must be made to assure that no more than the excess stock will be removed from each size class; this can be a tedious task for all but the most experienced foresters.

For whatever reasons, the selection method is currently largely unused on private lands in New England, although many guides exist to assist foresters in making the necessary assessments and estimations. The term "selection cutting" is often mistakenly applied to any kind of partial cut; see the discussion of selective harvesting in the Risky Treatments section below. What can legitimately be called selection harvesting is marking to leave optimum basal areas in an absolute minimum of three very broad size classes of poletimber, small sawtimber, and mature trees.

Artificial Regeneration

For a number of reasons, very little tree planting and direct seeding—*artificial regeneration*—are undertaken in timber management in New England. The main reason is that natural reforestation is rapid even after a severe disturbance. Also, the rocky and uneven terrain of the region is poorly suited to planting machinery. Finally, the considerable cost of planting or seeding must be carried through the full length of the relatively long rotations of New England species—an investment that many private landowners are unwilling or unable to make.

Planting involves acquiring seedlings (usually from 2 to 4 years old) that have been raised from seed in a nursery, and planting them on prepared sites. In direct seeding, seed is sown directly on a prepared site, bypassing the nursery phase. Seeding techniques include broadcast sowing, row seeding, and placing a number of seeds in small, scarified areas called *seed spots*.

For timber production in New England through artificial regeneration, planting softwoods is currently the most reliable financial investment. Success with planting hardwoods has been mixed. The introduction of protective seedling tubes and shelters which improve seedlings' growing

Table 17 Simplified Example of a Calculation for a Selection Harvest

DBH Class (inches)	Number of Trees per Acre	Optimum Number of Trees per Acre	Allowable Harvest of Trees
6–12	119	86	33
12–14	39	22	17
16+	21	12	9
All	179	120	59

Source: Adapted from W.B. Leak, Dale S. Solomon and Paul S. DeBald, *Silvicultural Guide for Northern Hardwood Types in the Northeast,* Research Paper NE-603, (Broomall, Pa.; U.S.D.A. Forest Service, 1989. The optimum goals used above are for average to above average sites.

conditions, has greatly increased survival rates—but has had corresponding effects on hardwood planting costs. White pine is the most frequently planted species, and larch and spruce are commonly planted in northern New England. Timber trees are usually planted to introduce a desired species that is not likely to occur through natural regeneration. A seed source may be absent, or unable to produce a crop in a given year when seed is needed. Planting or seeding may be the only ways to successfully convert a site dominated by a noncommercial or poorly producing species to a species capable of producing sawtimber. Planting softwood species under existing stands can be a cost-effective way to influence stand composition. Planting may also be undertaken on a badly burned area, where the recovery period necessary for the establishment of a valuable species would otherwise be several decades. Careful evaluation of the productivity potential of a site must be made to justify the high costs of planting. Seedlings can be obtained from commercial nurseries, and sometimes through state or county agencies. It is important to know the geographical origin of planting stock, since material from a warmer latitude may not perform well in all parts of New England.

Success with direct seeding has been mixed in New England, and the technique should still be considered experimental. The situations in which seeding might be used are identical to those for planting. White pine is the species most commonly seeded. Direct seeding has a dramatic cost advantage. The biggest problem involved is the protection of the seed from small mammals; if this problem could be solved, the amount of direct seeding in New England would probably increase. The geographical origin of seed is as important as the origin of planting stock. Seed is available from fewer sources than seedlings.

Artifical regeneration projects must be carefully planned and executed. Site characteristics, such as aspect, drainage, and soil type should be carefully assessed; the species selected must be suitable for the site. In most cases site preparation is critical to the success of planting and seeding, and it may represent a major portion of the cost of an artificial regeneration project. Preparation may involve the elimination of existing vegetation on a site and exposure of the soil in order to minimize vegetative competition with the introduced trees. In the case of direct seeding, exposure of the soil also enhances germination and root contact with the soil. Intensive methods of site preparation may involve controlled burning or the use of heavy equipment such as bulldozers or specially designed tractors. Herbicides are sometimes used in site preparation, but the controversy surrounding their use argues in favor of alternative methods such as mowing or scarifying whenever possible. Other methods of site preparation involve logging when the soil is not frozen or snowcovered. Since site preparation is uncommon in New England, homemade additions to bulldozers or skidders are generally used. On prepared sites planting by machine is possible, but in New England most planting is done by hand.

Follow-up inspections to evaluate the survival of planted trees are important. The first two years after planting or seeding are the most critical. It may be necessary to control competing vegetation mechanically (or chemically) for several years to give introduced trees a sufficient competitive advantage. A systematic estimate of seedling mortality should be made, followed by supplementary planting or seeding to compensate for seedlings that succumb or for poor seed germination.

References to readings on artificial regeneration are provided in appendix 2. In many counties, federal cost-sharing funds are available for seeding, planting and site preparation.

Risky Treatments

Some well-known and common cutting practices should be avoided. *Selective harvesting, commercial clearcutting,* and *diameter-limit cutting* are most often simply variations of highgrading.

Selective harvesting is not a silviculturally acceptable term, and it should not be confused with the selection method applied in uneven-aged stand management. It is a term which seems to be applied to any cutting in which some trees are left. In practice it is usually synonymous with the removal of the larger, more vigorous, but not necessarily older trees, after

Fig. 56. A commercial clearcut. The residual trees are of poor quality and vigor and have probably been damaged during logging. They will shade the site, which may inhibit desirable regeneration.

which a stand of smaller, weaker trees is left. Where such "selective" harvesting has been done in an even-aged stand, the residuals are often in the intermediate and suppressed crown classes of the original stand; they may be incapable of responding to release when the dominant trees have been cut, and are often of unwanted species. Selective harvesting is sometimes defended as a means of "letting the younger trees grow." Frequently, the smaller trees are not younger; they have simply been less vigorous, the losers in the competition for light, nutrients, and moisture.

In a true clearcut intended to regenerate intolerant species, all trees that are 2 inches and larger in diameter are removed or felled. In a *commercial clearcut*, only the merchantable material is removed in the harvest (fig. 56). Small, poor-quality, and marginal trees are left, often in a damaged condition. The residuals are likely to be unsightly, suppressed, of poor form, and of less valuable species. Residual hardwoods may develop epicormic branches, further lowering their quality. Finally, these residuals will shade the site and inhibit the intolerant species or advance regeneration that should be the focus of a clearcut. A logging contract for a clearcut should stipulate that all trees of 2 inches and greater in diameter must be removed or felled. Alternatively, an owner must be prepared to girdle, fell, or harvest for firewood what trees are not taken by the logger.

A *diameter-limit cut* is similar in effect to a selective harvest in that it often amounts to "taking the best and leaving the rest." Frequently loggers will contract to harvest all timber larger than a certain diameter—usually 10, 12, or 14 inches. Again, in an even-aged stand, this usually results in the removal of the high-quality, most vigorous trees. The lure of this method is the ease with which it can be applied and monitored, and the fact that it requires no consideration of the silvicultural needs of the stand or of the environment needed for regeneration.

Typical Sequences of Treatments

This section illustrates how a sequence of intermediate and regeneration treatments might be applied over the lifetime of a stand managed for sawtimber production. The discussion is general for the entire New England region; local factors, both biological and economic, create a tremendous variety of individual situations which need individual silvicultural responses. Also, the timing of intermediate treatments and regeneration harvests varies not only with the particular conditions of each stand, but with the needs of the landowners as well.

Hardwood stands.

In a 10-year-old hardwood stand, when the trees are from 1 to 2 inches in diameter, well-formed, vigorous trees of desired species are identified. A cleaning regulates spacing and quality and eliminates stems of unwanted species, which are clipped off and left on the ground. At age 20, the stand is about 30 feet tall, and has an average DBH of about 4 inches. A second cleaning removes defective, poorly formed trees and any remaining stems of unwanted species. The trees killed are eliminated by girdling or felling. Remaining are from 400 to 600 good trees per acre, space 8 to 10 feet apart. The tight spacing encourages good form in the crop trees. Enough crop trees remain to allow for inevitable mortality from natural causes. The trees killed are not removed from the site because they are too small to be useful even as firewood; it takes between 50 and 100 four-inch trees to yield a cord. After age 30, when the average diameter is about 6 inches, the stand is thinned at 6- to 10-year intervals. The least desirable trees are removed in each thinning, leaving trees at a proper basal area for good growth and the maintenance of tree quality.

Although optimal basal area increases as the trees grow in diameter, the number of trees per acre decreases.

Until this hardwood stand is about 50 years old, or has an average diameter of 10 to 12 inches, its products will probably be firewood, pulpwood, and a few small sawlogs, tie logs, and pallet logs. Where the stand is inaccessible, or the trees are unacceptable for fuel or pulp, unwanted trees continue to be girdled. When products are removed, roads are designed and built to serve current and future harvests.

After about age 50, a high percentage of the trees are large enough (12 inches in diameter and larger) to permit additional products from thinnings—sawlogs and perhaps veneer logs. Periodic thinnings continue, removing all but the best-quality trees to maintain the optimum basal area.

At about age 80, with many trees, 18 inches in diameter and larger, a managed hardwood stand might be ready for regeneration, and the remaining trees, those of highest quality, are harvested. If the two-cut shelterwood system is used, about half of the timber is removed at age 80. The resulting increase in light on the forest floor creates conditions favorable for the germination and establishment of advance hardwood regeneration. Within 10 years, but not until advance regeneration of acceptable species, size, and density is established, the remainder of the overstory is removed.

Softwood stands.

Management for softwood sawtimber production is strongly affected by where in the region it occurs. The principal softwoods in southern New England, white pine and hemlock, grow faster than the spruces and balsam fir which dominate northern New England. In general, established softwood stands produce merchantable volume faster than hardwood stands. In the seedling stage, however, softwoods tend to grow more slowly than hardwoods.

A cleaning at a diameter of 4 inches will remove stems of unwanted species with small crowns or poor form. These stands will usually be 20 to 40 years old, depending on dominant species and location within the region. At this point, the stand will have begun to differentiate into crown classes and potential crop trees can be identified. In the case of white pine, keeping a stand dense in the early years discourages the white pine weevil.

Following precommercial treatments in white pine, final crop trees can be identified and pruned.

The strength of local pulp markets will determine the point at which commercial thinnings can begin in a softwood stand. Where markets are weak, girdling may be necessary and commercial thinnings may not begin until a portion of the stand is sawlog size. In white pine stands in southern New England, sawlog thinnings in managed stands may begin at an age of 30 or 35 years. Growth of spruce/fir stands is significantly slower, thus delaying initial sawlog thinnings for 10 to 20 years. Later thinnings in a softwood stand will occur at about the same intervals as in a hardwood stand. The principal difference is that basal area will be maintained at a much higher level because softwoods are capable of growing well at much higher densities than are hardwoods.

Timber Management Techniques in Perspective

The applicability of the sawtimber management techniques just described to any given property depends on a number of factors, including the state of the woodland when acquired by a new owner, its potential, and the owner's priorities. Few owners pursue sawtimber management single-mindedly, and techniques are usually modified to the extent that other ownership objectives are accommodated. A written management plan provides the opportunity to allocate activities according to ownership goals.

Because of the land-use history of New England, many owners may have properties dominated by middle-aged and mature stands. One or more improvement cuts may be necessary to put the land into productive condition. Cuttings intended to improve a stand's sawtimber productivity will not have the desired result where stands are old and in such poor shape that the existing stand is not worth managing. Such stands should be regenerated. It is not always easy to recognize those stands that are beyond help; the assistance of a forester is recommended.

No matter how well designed, any sequence of treatments can be interrupted by events beyond human control, such as a major fire, an ice or wind storm, or a disease or insect epidemic. Management can reduce the risks from such dangers, but can never eliminate them. In the event of a major loss or alteration of a stand's condition, salvage treatments should recover whatever value is possible. Management objectives should be reviewed for the stand, and a decision must again be made as to whether the remaining stand is worth managing, or whether the stand must be regenerated.

For Fuelwood and Pulpwood Production

Fuelwood and pulpwood are usually byproducts of timber production, sugarbush development, wildlife habitat improvement, or clearing for recreational areas or trails. As such, they are the result of release or improvement cuttings, thinnings, or clearing operations, or of regeneration efforts where an existing stand is of low quality. In the management of private woodland, by far the most common generator of fuelwood and pulpwood is thinning operations.

Fuelwood thinnings might begin in a hardwood stand when it is about 30 years old and the trees average from 5 to 6 inches in diameter. Firewood will continue to be the product of thinnings until the stand is approximately 50 years old, at which time thinnings will begin to yield a good proportion of sawlogs. When firewood is a byproduct, and not the primary focus, of management, species favored in management will be determined by site conditions and local markets. However, it is worth noting that regeneration techniques favoring intolerant species (the seed tree method and clearcutting without advance regeneration) will produce stands in which firewood trees yield fewer BTUs per cord. On the average, intolerant trees have less dense wood than more tolerant trees; thus it takes a greater volume of wood from intolerant species to produce a given amount of heat when burned. Regeneration methods favoring species with medium to high shade tolerance (clearcutting with advanced regeneration and the shelterwood and selection methods) tend to favor stands containing species that have denser wood. Denser woods are preferred for firewood by people who cut their own wood and by commercial firewood operators. In areas where firewood markets are poor, but pulp is saleable and paid for by volume rather than weight, intolerant trees sold as pulp will be a better byproduct of thinnings, since more volume will be produced. However, it is important to keep in mind that when fuelwood or pulpwood is a byproduct of sawtimber management efforts, the regeneration method used will depend on the species best suited to sawtimber, not the best species for early thinnings.

In some cases, fuelwood may be the primary focus of management, rather than a byproduct. On very small properties (less than 10 acres) and on poor sites dominated by low-value hardwood sawtimber species (especially those that are capable of vigorous sprouting) fuelwood may be the logical or only choice for management. For residential fuel, management will attempt to maximize the heat value of each cord produced. Hardwood species with high BTU values per unit volume of wood, such as oak and beech, are most desirable. With the exception of liberation cuttings to eliminate large old culls, release cuttings are best foregone until the stand is beyond the sapling stage, so that the trees to be removed will be large enough to use as fuel. Thinnings attempt to capture the volume of those trees that might die from suppression, and should be just light enough to keep a site fully occupied; increasing the diameter growth on individual trees is not the objective.

In fuelwood management the selection of crop trees is based on species and vigor only. Form is of little consequence, except that it is easier to split and stack straight trees. Stands managed for a primary product of firewood should be regenerated relatively early, generally when they are not beyond 12 or 14 inches in diameter. Fuelwood of a larger size is hard to handle, and sprouting ability diminishes as trees grow beyond this diameter range. Winter harvests encourage prolific sprouting in those species with the capacity to sprout. Trees which originated as stump sprouts are especially efficient fuelwood producers because they grow more quickly than seedlings; many trees of fuelwood size are produced from a single stump. If sprouting is the principal source of reproduction, the regeneration method used in fuelwood management is not important as long as enough light reaches the ground to allow good growth of the sprouts. If regeneration from seed is desired, regeneration techniques will be similar to those used in timber management of the same species.

On poorly drained sites with low timber-growing potential, pulpwood is occasionally a logical primary product of management. Stands of balsam fir, a species that often begins to deteriorate before reaching sawlog size, are sometimes suitable for pulpwood management. In these cases, it is usually not necessary or economically efficient to apply any intermediate treatments. Regeneration is usually best accomplished by clearcutting, but it should be noted that such sites are likely deer yards.

An intensive approach to fuelwood and pulpwood management, still experimental in New England, is the cultivation of planted stands for maximum production of wood fiber per acre per year. Some very fast-growing species, such as silver maple, aspen and other poplars, black locust, ash, alder, red pine, and larch, are capable of growing much more wood fiber,

Table 18 Estimated BTUs of Selected Fast-Growing Species

	Millions of BTUs per Cord	Millions of BTUs per Acre per Year
Alder	27.2	40.8
Ash	29.7	38.6
Aspen	18.2	40.1
Beech	33.2	36.6
Birch	29.7	14.9
Cherry	25.0	32.5
Cottonwood	18.7	52.3
Hickory	40.4	32.3
Larch	23.6	26.0
Locust	37.3	48.5
Maple	31.3	47.0
Oak	37.5	37.5
Hybrid poplar	18.7	59.8
Sycamore	23.5	47.0
Walnut	24.9	10.0
Willow	17.7	53.0
Yellow poplar	19.4	25.2

Source: C. M. Hunt, U.S.D.A. Forest Service, State and Private Forestry (Broomall, Pa.), personal communication, 1975.

or *biomass*, per acre annually than are traditional pulpwood and fuelwood species (see table 18). A cord of aspen fuelwood, for instance, produces less fiber and fewer BTUs than a cord of oak because it has less dense wood; but an acre of land can grow more tons of aspen than of oak over the same period of time, thus more than compensating for the lower weight per unit of wood volume. Such experiments with biomass plantations are usually aimed largely at the commercial markets, for which plantation trees are clearcut at a young age and ground into chips for boiler fuel or for papermaking.

Intensive biomass production may never be popular on private land in New England. Especially on smaller holdings, such production might conflict with aesthetic values, although it does offer the advantage of yielding more fiber on less acreage than traditional methods. The approach is limited to relatively level sites where cultivating and mowing equipment can operate. It is doubtful that such intensive production, in which trees are harvested at very short intervals, could be sustained over the long term on New England's relatively infertile soils without the great expense of fertilization. Considering the costs of cultivation, mowing, and possible fertilization, it seems unlikely that biomass plantations offer any economic advantage over traditional fuelwood management for New England landowners.

For Maple Sap Production

The techniques used in managing an even-aged sugarbush are quite similar to those involved in even-aged sawtimber management in that they may involve release cuttings, improvement cuttings, thinnings, and eventual overstory removal. Release cuttings and improvement cuttings in previously unmanaged stands will serve one purpose: to exclusively favor sugar maple. Conifers should be eliminated from a stand since they are thought to diminish sap production by competing with the crop trees for moisture, slow the warming of the forest floor in spring, and delay sap production by shading tube lines and tapholes, thus preventing them from thawing.

Stem form is not important in sugarbush management; the development of a full, large, well-proportioned crown is the major goal of early intermedite treatment in a sugarbush. As earlier chapters have pointed out, the optimum tree density (basal area) per acre is much lower for sugarbush development than for sawtimber production. Fewer trees spaced widely will result in broad, deep crowns, capable of producing more sap with a higher sugar content. Improvement cutting is not warranted unless the sugar maple trees in the stand can produce sap of average sweetness and are capable of vigorous, prolonged growth following cutting.

Thinnings in a sugarbush are designed to increase crown size, tree vigor, and diameter growth rate. Rapid diameter growth allows a tree to reach quickly the minimum size for tapping. Selections for thinning among sugar maples whose crowns are crowding each other should be made on the basis of tree characteristics and relative sap sweetness. Sweetness is determined with a refractometer, which measures the percentage of sugar content in sap (see chap. 2). Trees can be tested in the spring or fall by obtaining a few drops of sap from small holes made in the trunk. Trees with the highest sugar content are selected for retention. Although sweetness levels vary annually, the relative ranking of trees in a stand does not seem to change; the sweetest trees remain so.

A sugarbush will eventually deteriorate in health and productivity, although the stand may be 150 years old or more before it does. Because of the substantial shade tolerance of its seedlings, sugar maple can be regenerated in even-aged stands by the shelterwood method. However, after the final removal of the large trees a sugarbush will become unproductive for many years before the next generation reaches the 10-inch diameter at which it can be tapped. Tapping trees before they attain that size is likely to seriously reduce their vigor.

An alternative to the shelterwood method is the creation of an uneven-aged stand by applying the selection system. Continuous sap production would be maintained, although untappable trees would always be present. While it is a challenging undertaking, the selection system may be easier to apply in sugarbush management than in timber production. Detailed records are necessary in sugarbush management, and tree density is lower, so maintaining an inventory of diameter distribution may be less burdensome.

Two types of experimental techniques are of current interest. The first is low-nitrogen fertilization for improving the growth rate and health of a sugarbush. The results to date are not sufficiently promising to warrant the significant cost of such fertilization. The second technique involves the selective breeding of "sweet" trees. Since it appears that the relative sweetness of sap is partly a result of genetic factors, researchers have been attempting to develop improved sugar maple stock that can be planted under old sugarbushes or on open land. These trees are available although sugarbush plantations are rare, surely due to the significant waiting period between planting and production.

For Wildlife Habitat Improvement

Two dangerous myths often influence the consideration of wildlife habitat management techniques: according to the first, the cutting of trees threatens wildlife; according to the second, all forest cutting benefits wildlife. As usual, the truth is not as simple as the myths. Cutting, where it creates diversity in relatively uniform expanses of forest such as typify areas of northern New England, will encourage many species of New England wildlife if it is carefully planned and executed. Second, although harvesting may benefit many *more* species than does unbroken forest, it may adversely affect a few others, which are likely to be the most rare or retiring. Finally, it must be remembered that, in some areas of the region, especially southern New England, unbroken forest, which is an important habitat element, is increasingly rare and should be conserved.

Wildlife habitat management can be intensive—aimed at improving conditions for a specific animal—or extensive, with the goal of providing habitat for as many species as possible on an ownership. The choice of forest management techniques for intensive habitat management is entirely dependent on the species of interest, and cannot be adequately discussed here. In seeking reference materials, a landowner will probably find that detailed information is abundant on habitat management for game species, but that similar information is scarce for nongame animals. However, because intensive management for a single species is likely to eclipse other forest values, and because many landowners are interested in the nonhunting values of wildlife, the encouragement of maximum numbers of both game and nongame species is the common purpose of habitat improvement on private lands.

Each wildlife species seeks a special habitat, a certain arrangement of food, cover, and water. Techniques carefully designed to blend and link together many different habitats, each of which may benefit several species will increase wildlife diversity. Linking habitat elements with corridors of undisturbed forest is essential to their usefulness to many species reluctant or unable to cross open, disturbed or developed areas to reach them.

Cleanings are not widely applicable in habitat management, except to insure that mast-producing species are not eliminated from a stand by competition in the early stages of development. To release a stand that is valuable to wildlife, liberation cuttings can be applied, and overtopping trees that are 8 inches or larger in diameter can be girdled, rather than cut, to yield a double benefit. The snags created by girdling are crucial sources of food and cover for many birds. Very large, hollow snags provide dens for several forest mammals. The release of old apple trees by removing vegetation above and around them is necessary if they are to be productive sources of wildlife food.

In stands with unbroken forest canopies there is virtually no understory, which is an important habitat component for many species. Treatment that opens up some of the canopy and allows sunlight to reach the forest floor, such as improvement cuttings and thinnings, can be used to diversify habitat. A diversity of songbirds is greatly encouraged by cutting that results in vegetation of three distinct heights: a layer less than 2 feet tall, a second layer from 3 to 25 feet in height, and a third layer taller than 25 feet, each of which provides habitat for certain woodland birds. An herbaceous layer of ground vegetation can be established early in the life of a stand by doing a light thinning as soon as possible. Treatments in which less than 20 percent of the basal area is removed will promote the development of an herbaceous layer. Very heavy cuttings that remove from 40 to 50 percent of the basal area tend to favor the development of a woody understory of shrubs and small trees. If the woody understory becomes well established, it may completely shade and eliminate the herbaceous layer. Therefore it is better to make lighter, more frequent thinnings if a layer of vegetation less than 2 feet in height is desirable.

Special attention should be given to overstory mast trees during thinnings. By creating ample space around them—particularly by removing those trees offering crown competition—the production of greater quantities of fruit and nuts is encouraged. Very productive mast trees often warrant intensive care to maintain their vigor and crown size. Thinnings made to aid mast trees will also promote desirable understory development.

Treatments for habitat improvement should leave about five snags per acre. In hardwood and mixed stands, sprouts from cut trees will provide needed food for browsers such as deer. Brush piles constructed from the slash left from thinnings are very effective as shelter for a number of forest mammals. When seeded with a mixture of grasses, legumes, and other herbaceous plants, roads and trails built for access to a stand can be effective as cover and as sources of insects eaten by birds. Making roads and trails a few feet wider than necessary will further improve habitat.

Dense stands of softwoods that are used for shelter by deer in the winter should be cut lightly, if at all. Snow depth under the canopy of these stands is often only half what it is under hardwoods. Species such as hemlock, pine and spruce intercept the snow very effectively, and provide a good deal of protection from heavy winds as well. Too heavy a cutting in softwoods can virtually negate the value of a stand as a winter deer yard. Very light cuttings may increase the value of the stand by encouraging small amounts of regeneration. It is better to leave a softwood stand uncut than to cut it too heavily, because deer use such stands primarily for cover. However, a nearby source of food is also critical; one beneficial practice is to clear a strip from 25 to 75 feet wide around the edge of a softwood stand to provide browse near primary cover.

Pruning is occasionally applicable in habitat improvement. It is most commonly used to maintain the vigor and decrease the risk of wood decay of old apple trees and other mast producers. Pruning live branches, however, may increase the size of the fruit without increasing a tree's gross annual production, and thus is of little benefit to wildlife. Fertilization of an individual apple tree may be necessary to maintain the tree's vigor and value to wildlife.

Regeneration harvests staggered over time will result in a mixture of species, age, and size classes on a property. For habitat improvement purposes, a guiding principle of such harvests should be the maximization of the length of borders between different stands, between woods and open land, and between water and upland areas. These borders, or edges, are where wildlife thrive because they are the places where habitat elements converge. For instance, many species of songbird congregate at field and forest edges. The fields, which host much insect life, provide food, and the forests provide cover. Deer will frequent the borders between mature and sapling stands, since one provides cover and the other browse.

Each regeneration method has applicability to wildlife habitat improvement. Clearcuts (and seed tree cuts) create a dense mix of vegetation at ground level, providing prodigious amounts of food and cover for many species. Aspen, a species favored by clearcutting, is an extremely productive tree species for wildlife purposes (fig. 57). Clearcuts can be laid out in strips or in irregularly shaped patches to maximize edge. The division of a woodland parcel into several patches is a means of increasing the diversity of age and size classes among trees. At 10-year intervals, one-tenth of the patches should be regenerated in a pattern such as that shown in figure 58.

The shelterwood method provides periodic renewal of understory food sources while maintaining a degree of high cover. With this method the total disruption of habitat is less sudden than it is with the clearcut or seed tree methods. The selection method is very well suited to the maintenance of stable conditions and the vertical layers of vegetation that provide a desirable diversity of habitat elements.

On properties where only even-aged harvesting methods are used, wildlife benefits if from 5 to 10 percent of the woodland is left unmanaged to ensure the availability of some mature forest conditions. Such an area should not be selected simply because it is unsuitable for other uses. It can be tempting to relegate wildlife to those areas having poor sites for growing timber, but these sites are often poor areas for wildlife as well. *Good land is usually good land for all uses; if a stand is too infertile to support good timber, the quality of the food it produces for wildlife is also likely to be low.*

Artificial regeneration methods aimed solely at habitat improvement are very expensive. The planting of shrubs and trees creates rapid regenera-

Fig. 57. Aspen regeneration
9 months after a clearcut for
the purpose of improving
habitat for ruffed grouse.
The numbers on the staff indicate height in feet.

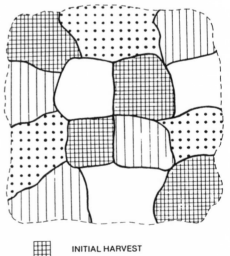

Fig. 58. A simple cutting
plan to improve wildlife
habitat by increasing age
and species diversity and
edge. Edge could be further
increased by making the
borders of the cutting areas
as irregular as possible.

INITIAL HARVEST

HARVEST — 10 YEARS LATER

HARVEST — 20 YEARS LATER

HARVEST — 30 YEARS LATER

tion but it is rarely affordable on a large scale. Large planting projects for wildlife should be undertaken only after attempts to work with the existing vegetation have proved inadequate.

In efforts to improve forest wildlife habitat, it is important to remember that all wildlife species need an accessible supply of water. For some species, surface water covering a substantial area is necessary; this can be provided by importing beavers to a stream on the land, or by excavating an area where there is a seasonal or year-round spring. Beavers should not be introduced without consultation with a state fish and game department, and their impact on neighbors' lands must be taken into consideration. Beavers breed prolifically, and may end up colonizing all streams in a given area. Wildlife ponds (excluding fish ponds) can be quite shallow; they usually do not need to be more than 4 feet deep. Both the state fish and wildlife department and the Soil Conservation Service should be consulted about the construction of wildlife ponds and management of beavers (see appendix 4).

It is worth emphasizing that feeding and releasing wild animals and controlling predators are practices that are not appropriate to wildlife management on private woodlands. By feeding animals a landowner can create or maintain an artificially high population that the natural habitat of an area is unable to support. Feeding deer in winter can be fatal to them. Released animals will not survive or remain in an area if the habitat is not suitable. Predator control is usually either ineffective or very upsetting to the ecological balance of an area, and often creates more severe problems than it seeks to solve. Habitat improvement and maintenance are the most consistently beneficial and effective tools for wildlife management on private land. If suitable habitat for a species exists, and members of the species live nearby, the animal will move into the habitat.

For Recreation and Aesthetics

Specifying techniques to improve the aesthetic appearance of a stand of trees is difficult, and depends entirely on the tastes of the owner. Common aesthetic improvements include opening vistas, installing roads or trails, pruning, and increasing the diversity of the vegetation.

Vistas can be created simply by removing the vegetation that blocks a view. A narrow corridor, or "rifle shot," can open a view while also creating edge that is desirable for wildlife. Particularly interesting trees or groups of trees can be highlighted by clearing a sightline to them. Regrowth of an understory in corridors and sightlines will be rapid, especially in hardwood stands. An owner must be prepared to reenter the area to keep the height of vegetation down.

Roads and trails often enhance the aesthetic appeal and recreational potential of a property, making it easier to walk, ski, or snowmobile through the woods. A loop trail that goes near or through vegetation of all types and ages and includes different elevations is very effective. If possible, a recreational trail should have an attractive midpoint, such as a vista or a stream. On many sites recreational trails can also be used for the removal of fuelwood and fire protection.

Pruning can improve a short view into a forest, and in dense softwood stands makes walking through the woods much easier. In hardwood or mixed stands, increasing vegetational diversity will allow for a more varied and impressive display of fall color. Maintenance of some softwoods in a predominantly hardwood stand adds a pleasing touch of green during the dormant season. An herbaceous layer of ground vegetation, favored by light thinnings, will produce spring and fall flowers.

Many people like the look of open, parklike stands. This effect can be achieved by thinning. However, the technique can backfire, leading to such a proliferation of sprouts, shrubs, and ground vegetation that views through the forest are obstructed and navigation is impeded. Maintaining a closed canopy is the best way to keep the understory sparse. Low thinnings that do not alter the upper canopy are recommended.

The shelterwood and selection methods are probably the regeneration techniques that are best suited to most landowners' aesthetic tastes. Change is gradual, and when the selection method is used may be sometimes hardly noticeable. However, in hilly country a properly applied clearcut can produce striking views of the surrounding countryside. Such a cut should be laid out with irregular edges following natural contours; square or rectangular clearcuts are visually jarring. Buffer strips of vegetation must be left along all streams. Release cuts are rarely needed in management for aesthetics, except perhaps when an owner wishes to release paper birch from overtopping by other, less aesthetically pleasing species.

Some Final Considerations

The applications of management techniques for several woodland uses have been described in this chapter. Because each use has been dealt with separately, it is necessary to emphasize again that it is common practice to combine uses—timber management, fuelwood production, and recreation, for example—in a single stand. A few uses are incompatible, but most combinations of uses are feasible (see chap. 2).

Achieving most woodland objectives requires some form of intermediate treatment. These treatments often represent a cost to the owner: a financial outlay is necessary to alter stand conditions to the desired state. Depending on demand for fuelwood and pulpwood, the prospects for economic returns from most intermediate treatments are variable.

With regard to regeneration, the shelterwood method of harvesting and regeneration may currently be the closest thing to a "right" method for private forest landowners in New England *when and where it is suitable*. The overstory is not removed until a new forest is safely in place, and the results of the technique are aesthetically pleasing, provide a continuous forest cover, and are compatible with recreational use of the land. Ideally, shelterwood will stimulate the production of both tree seed and woody browse, so it is useful for the production of wildlife food as well. In addition, the method is suitable for New England's even-aged forest; it is not as complicated as the selection method; and it is appropriate for the many New England species that tend to be dependent on starting as advance regeneration.

Finally, it is important to make two points about the management techniques described in this chapter. First, they are no more than tools for accomplishing a desired objective; they have no intrinsic value, good or bad. Clearcutting, for example, is a sound forest practice when properly applied, although the silvicultural conditions in which clearcutting is applicable are quite narrow. They exist when specific wildlife habitat requirements call for increased areas of "edge," brushy growth or intolerant tree species such as aspen; second, when advanced timber regeneration is present and ready for release; and third, in the rare case of a stand all potential value of which has been eliminated by natural catastrophe or past practices. Clearcutting is not always undesirable and selection cutting, often thought to be inherently good, can be disastrous if misunderstood and misapplied.

Second, while success in using the techniques depends on skill and timing, chance always plays a role. Application of a technique "by the book" does not guarantee success. Seed production in a desired species may fail repeatedly because of an insect over which the forest manager has no control; a wind storm may flatten a stand that has been carefully tended for years. Local conditions are important: the same treatment applied to similar forest stands in different locations may have widely varying results. For example, a high population of deer in an area may foil the most well-designed attempts to regenerate desired hardwood species. We are only one influence in the forest; it is a mistake to think we control it.

6   Harvesting Forest Products

Logging operations are a critical element in forest management for any objective, and the choices made in planning and executing them have ecological, visual, and financial implications for many years. Although a poor logging job may maximize immediate financial gain, years of cultural effort and future productivity can be negated if the logging is poorly conceived or negligently executed. A good logging job will show sensitivity to possible visual changes, leave the forest vigorous, and still generate income. With planning, logging can also result in a useful and durable system of roads and trails and in improved wildlife habitat.

Negative experiences with logging and unsatisfactory dealings with loggers are the usual causes of a landowner's resistance to forest management. Averting these situations depends greatly on an understanding of the processes of selling trees, cutting them, and extracting them from the forest; it also depends on the owner's (or the forester's) paying careful attention to four other aspects of harvesting: choosing a buyer for the trees to be sold, designing a road system, composing a sale contract, and evaluating the logging job.

In the special case of maple sap, harvesting operations involve thinning the sugarbush, tapping trees, gathering sap, and performing a variety of maintenance operations in a stand. A landowner who leases a sugarbush to others must deal with most of the same general concerns as a landowner who contracts for a logging job, such as finding a reliable buyer, determining access, drawing up contracts, and evaluating woodswork. The sugarbush owner must also deal with some concerns that are unique to maple sugaring.

Selling and Harvesting Trees

Types of Sales

Trees can be sold for wood products according to any of several arrangements, each of which has advantages and disadvantages. Most sales are of

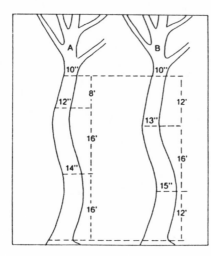

Fig. 59. Two ways of cutting the same tree into logs. *Left:* In scheme A, the three logs yield 182 board feet. *Right:* In scheme B, the three logs yield 275 board feet and 42 percent more revenue. In scheme B, less wood was wasted at the sawmill because the emphasis was on maximizing log straightness, not log length. The efficiency with which trees are cut into logs greatly affects the actual wood volume realized in a lump-sum sale. More important is the effect that this efficiency has on the seller's income from a unit-priced sale.

stumpage—trees as they stand on the stump. Trees to be cut should be designated with two painted marks: one at eye level and one at the base. The first is easily seen by the logger, and the second serves as assurance to the owner that only the designated trees were removed during logging operations.

The two most common methods of selling stumpage are *lump-sum* and *unit-priced* sales. In a *lump-sum sale* of sawtimber all designated trees are sold for a single sum, set either by bid or negotiation, that is based on the owner's or the forester's estimate of the volume of merchantable wood they contain. It is important to understand that volume is determined (scaled) from the measurement of tree diameter and an estimate of the saleable portion of each tree's length—that is, the portion below a fork in the main stem, or below the point at which the stem becomes too small for use in the desired product. Volume of designated trees can be determined by the measurement of all trees (100% tally) or by a sampling procedure. The higher the value of the timber, the more trees are sampled. The logs that later come from the tree may have a different total volume, based on how much of the tree is used and how efficiently it is cut into logs (fig. 59). It is the volume of the tree as estimated on the stump that is tallied in a lump-sum sale.

For sawtimber or veneer products, volume estimates are given in terms of board feet, and are taken from a table called a log rule that shows the estimated amount of lumber in a log or tree of a given diameter and height (or length). There is more than one log rule; each is based on a different assumption about how a log would be sawed, and how many board feet of lumber would result. Four log rules are compared in table 19. The use of log rules is an archaic practice that persists in spite of widespread agreement that there must be a better way. In most New England states, any log rule is legal for scaling trees or logs for a sale, as long as there is a specific agreement on a rule between buyer and seller. The volumes of fuelwood, pulpwood, and chips are given in cords, cubic feet, cunits, or tons, depending on which measures are legal in a given state (see appendix 3).

In a lump-sum sale, the estimated volume of wood is not guaranteed. Depending on the accuracy of the owner's or the forester's inventory, the extent of internal defects in the trees, and the buyer's skill in maximizing the use of each tree, the actual volume will probably be more or less than the estimate. In either case, no price adjustment is made; it is the designated trees that are sold, not the volume of wood. For the seller, a major advantage of a lump-sum sale is the incentive for the buyer to use as much of each tree as possible. If the buyer has paid for the whole tree, he or she will be much more likely to squeeze an additional 8-foot log out of the top. The use of trees to the fullest extent possible reduces the amount of unsightly debris left in the woods, and in the case of softwoods reduces the hazard of fire following logging. Another advantage of the lump-sum sale for the seller is that no trees are cut unless and until they are paid for; the buyer always pays ahead.

In a *unit-priced* (or *pay-as-you-go*) sale, designated trees are measured after they are cut and are paid for based on the measurement of board feet

Table 19 Comparison of Four Log Rules

Small-End Log Diameter (inches)	Board Feet in a 16-Foot Log			
	International ¼-Inch Rule	Doyle Rule	Maine Rule	Vermont Rule
10	65	36	68	64
14	135	100	142	126
18	230	196	232	210
22	355	324	363	315
26	500	484	507	342
30	675	676	706	590

Note: The International ¼-Inch Rule is the one most widely used in New England.

or cords they are estimated to contain. A price per thousand board feet (abbreviated MBF) or cord is quoted, and the buyer pays according to the amount of wood received. A unit price for sawtimber is set for each species of tree designated for sale, and, for simplicity, is always given per thousand board feet. Occasionally unit prices are set for each grade of log quality for each species, although the practice is uncommon in most of the region. In most New England sales sawtimber is sold "woods run"—that is, priced according to an estimated average value of the marked trees of each species.

Each species of softwood pulp available in a sale is usually named, but all hardwoods are usually priced together as "hardwood pulp." Unit prices for pulpwood and fuelwood are per standard cord ($4' \times 4' \times 8'$), cubic foot or per ton. Scaling by weight is an alternative method, but it is illegal for selling fuelwood in some New England states (see appendix 3). Weight is not a good indicator of the value or quantity of fuelwood, unless it is combined with a measure of moisture content.

The amount of wood removed in a unit-priced sawtimber or pulpwood sale is estimated by measuring logs either at the landing before they are loaded onto a truck or, more commonly, at the mill. When the logs are measured at the mill, a unit-priced sale is called a *mill-tally* sale. In either case, the landowner must trust the honesty of whoever scales the logs. When volume is based on a mill tally, the landowner should receive copies of all tally sheets as receipts. Unless the volume in marked trees is in some way estimated at the time of marking, there will be no way for the owner to confirm that the tally total is honest. Most foresters make some form of volume determination of marked trees even if payment is to be made on a unit-priced basis.

For the seller, a disadvantage of the unit-priced sale is the lack of incentive for the buyer to utilize a tree to its fullest extent, since the buyer pays for only what leaves the site. This problem is highlighted in figure 59.

Combinations of lump-sum selling and unit pricing are uncommon but possible, and they often eliminate some of the drawbacks of each arrangement. An example would be an arrangement by which the buyer agreed to purchase the estimated volume of the designated trees (as in a lump-sum sale) at rates based on the grade of the logs determined by the mill.

Each method of selling stumpage has advantages and drawbacks. Lump-sum sales imply the involvement of a forester, at least to the extent that the trees to be cut have to be marked and measured. Lump-sum sales are advantageous because a landowner knows when the contract is signed what the revenues from the sale will be. Lump-sum selling assures that all buyers are bidding on the same trees. It avoids the problem of having loggers say that they will do what is "right" for the woodlot, then discovering that their perceptions of what is right differ from one another's and from the landowner's. Lump-sum sales commonly involve one or a small number of payments in advance for the timber. A lump-sum contract with a stiff payment schedule may eliminate small operators who are capable of paying a decent price and doing good work, but who are unable to tie up large amounts of cash for the entire period of a cutting contract. Such operators are better able to compete in a unit-price situation. Both types of sales may qualify for capital gains treatment under federal income tax laws if the timber qualifies as a capital asset, although a unit-priced sale is less likely to be questioned than a lump-sum sale. (Federal taxation of timber is discussed in more detail in chapter 7.)

An alternative way of selling forest products is to market logs rather than trees—to sell "roadside." An owner can cut trees into logs and move them to a roadside, or he or she can pay a logger—usually by the board foot or cord—to do so. (Logging is very dangerous and requires considerable skill; most landowners would be well advised to leave logging to professionals.) The logs are then sorted according to their product potential—sawtimber, veneer, pulpwood, or fuelwood—and sold to a buyer who will truck them away. Because value is added to the logs by investing the time and money to move them to the roadside, such an arrangement may increase the financial return from a sale, as well as allowing some discretion in the allocation of costs to logging and stumpage. This flexibility permits an owner to report a smaller timber sale profit to the Internal Revenue Service (IRS), which in turn lowers income tax liability. Landowners whose properties are located close together may be able to jointly market high quality logs "roadside." This arrangement can work when no individual alone has a sufficient volume of a particular species or grade to make a full truckload.

In any case, only the most experienced landowners should attempt to sell logs roadside without the assistance of a forester. Before the trees to be sold roadside are cut, it is absolutely critical to have secured a buyer, to understand the buyer's log specifications, and to be assured in writing that the buyer will take and pay for the logs promptly after they are cut. Logs cannot be kept indefinitely; decay, blue stain fungus, and insect damage may begin in some softwoods within a few days after cutting in summer; hardwoods store a little longer. The sale agreement should also make clear to both parties the buyer's specifications for log length, diameter, and quality.

In some cases, a landowner may be able to arrange a harvest without the exchange of money, receiving instead lumber or fuelwood for personal use. The buyer will then harvest the trees, and perhaps saw and return a portion of the lumber; the returned proportion is usually about 1 board foot of lumber delivered for each 5 board feet of stumpage harvested. In a barter arrangement where the landowner cuts the trees, moves them to a landing and has the logs sawn on-site by the operator of a small portable sawmill, about half the lumber will be returned. Landowners interested in obtaining a residential fuelwood supply from their woodlots can contract with a local fuelwood cutter, or make arrangements on a share basis. Under a contract arrangement, the cutter will harvest designated trees and deliver them to a location of the landowner's choosing. The landowner pays the cutter by the cord. Under an agreement to share, there is no exchange of money. The cutter receives an agreed-upon number of cords for each cord delivered to the landowner; a ratio of 3 or 4 cords of stumpage in exchange for each cord delivered to the owner is common. The landowner and the cutter should sign an adequate contract covering the terms and conditions of the arrangement. Most such jobs involve small, low-impact equipment, so liability is more of a concern than damage.

A landowner inexperienced in selling timber would be wise to sell on a lump-sum basis. It affords the most protection, while minimizing the landowner's responsibility for matters he or she knows little about. Unit-priced sales and log marketing are recommended only when an owner has a well-developed, trusting relationship with a buyer.

Locating a Buyer

In New England, most buyers are either independent loggers or mills that employ logging crews on a full-time or contractual basis. Independent loggers own their own equipment and purchase stumpage directly from landowners, or through foresters who act as landowners' agents. They harvest standing trees, and market the various products. Logs are sold to sawmills or veneer mills; pulpwood to a pulp mill for papermaking; fuelwood to homeowners, pulp mills or biomass energy facilities. Sawmills and paper mills that purchase stumpage may have woodsworkers among their employees. These company loggers operate company-owned equipment and are paid by the mill, usually by the hour.

Landowners, and purchasers of standing trees who do not employ full-time woods crews, may retain loggers on contract to do their harvesting on a piecework basis. Like independents, contract loggers own their own businesses and equipment. Regardless of the time it takes to harvest the timber, they are paid on a unit basis for the material harvested. A contract logger, unlike an independent, does not own the harvested trees. An independent logger can make a profit through skillful marketing of harvested material; a contract logger, on the other hand, can profit only by working efficiently in the woods.

The price offered to the seller is clearly a primary consideration when choosing a buyer. Prices are set in one of two ways: by negotiation between the buyer and the seller, or by bidding. After the sale trees have been marked and scaled, bids are solicited by sending a prospectus to potential buyers who are known to be responsible loggers. The prospectus describes the trees, their location, and some of the conditions of sale which will be set forth in a contract; it specifies a date when the owner or a forester will show the trees to interested parties (fig. 60). After the showing, sealed bids are accepted until a specified date (fig. 61). The owner decides which bid, if any, to accept. The bidding process is advantageous because it offers the seller the best price available, and provides a clear, objective measure of the current market value of the trees.

Price should not be the only factor in the choice of a buyer. A buyer's reputation for careful woodswork and business dealings should also be considered. A sale negotiated directly with a known buyer, bypassing the bidding process, may be the best option available to a landowner if the buyer's reputation is very good and the price negotiated is known to be fair. While the bidding process assures the best available price, it does not guarantee the desired quality of work during harvesting operations. Even within the bounds of a contract, there may be a wide range of acceptable logging practices. In a negotiated sale, a buyer can be selected who is willing to do the type of job the owner desires. However, unless an owner is familiar with markets and market conditions, the evaluation of any offers will be difficult, so the inclusion of a forester in the negotiating process is recommended. In assessing the differences between negotiated prices and anticipated bid prices, the owner must take into account any differences in work quality that may be expected. Negotiated sales work best when the owner or the forester is confident of the buyer's ability and willingness to do the job as requested.

FOREST CONSULTANTS, INC.

Box 321, RFD 3
Clifton, CT 06999

NOTICE OF INVITATION TO BID ON FOREST PRODUCTS

—Forest Consultants, Inc. are offering for sale an estimated 137,385 board feet of timber and 84 cords of pulpwood/fuelwood. A detailed tally is attached. Trees to be sold have been marked with blue paint. The trees marked for cutting will be sold on a lump-sum basis, using the International 1/4" Rule. Timber has been marked on approximately 40 acres.

TIMBER DESCRIPTION

Species	Estimated Volume	No. of Trees	Avg. Vol/Tree
White pine	73,550 BF	267	275 BF
Red oak	21,420 BF	102	210 BF
Black birch	18,130 BF	130	139 BF
Red maple	24,285 BF	221	110 BF
Hardwood fuelwood	84 cords	504	.17 CD

The timber is located on the Adler property in the town of Laneville. The timber will be shown January 24, 1994 at 8 a.m. Interested buyers should meet at the Laneville Post Office . See attached map.

—Sealed bids should be sent to the address listed above. Bids should contain the lump sum amount offered for the marked timber. Bid envelopes should be marked "Bid—Adler" and will be accepted until 4 p.m. on February 9, 1994 , at which time they will be opened and read. All bidders will be notified of the results within one week. The landowner reserves the right to reject any and all bids.

—The successful bidder will be expected to sign the timber sale contract within 10 days. Full sample contracts will be available at the showing. The following payment schedule will be required:

* 20% at the time of contract signing;
* 30% before the beginning of operations, or by August 1, 1994, whichever comes first;
* 50% within one week after the beginning of operations, or by October 1, 1994, whichever comes first.

—A performance deposit of $1000 will be required.

—For additional information, call (203) 999-1234.

Fig. 60. A sample sale prospectus. There is no standard form for a prospectus. Any special requirements or conditions associated with a sale will usually be pointed out in the prospectus.

```
┌─────────────────────────────────────────────────────────────────┐
│                          ──────────                               │
│                          BID SHEET                                │
│                          ──────────                               │
│                                                                   │
│   I/We agree to purchase the marked standing trees described in   │
│   Forest Consultants' Notice of Invitation to Bid:  Adler  for the total │
│   sum of $_____ , according to the payment schedule described in │
│   the Notice. I/We agree also to furnish, prior to the beginning of │
│   logging operations, a performance deposit in the amount of      │
│   $ 1,000   to be held in escrow. The breakdown of my/our bid is as │
│   follows:                                                        │
│                                                                   │
│             Species                                               │
│        _____   $_____ /MBF        │
│        _____   $_____ /MBF        │
│        _____   $_____ /MBF        │
│        _____   $_____ /MBF        │
│        Fuelwood                    $_____ /cord        │
│        Pulpwood                    $_____ /cord        │
│        Biomass                     $_____ /ton         │
│        Firm name: _____ │
│        Address: _____ │
│                 _____ │
│        Telephone: _____ │
│        Signature: _____ │
│        Date: _____ │
│                                                                   │
└─────────────────────────────────────────────────────────────────┘
```

Fig. 61. A bid sheet. This form is given to prospective buyers with the sale prospectus, and returned to the forester (or seller) by those interested in bidding.

A list of loggers can be obtained from an Extension Service office, from a county forester, or from other landowners who have sold timber and may be able to suggest one or more reputable operators. Obtaining references from a potential buyer's clients and visiting the buyer's recent logging jobs are recommended. A landowner should discuss all expectations and concerns with each prospective buyer, or with the forester who acts as the owner's agent. It is critical that a logger be made to understand the silvicultural purposes behind the design of any harvest: what species are being grown or regenerated, what techniques are to be used to encourage them, and what measures are being taken for the protection of soil, water, aesthetics, and wildlife habitat. Most loggers will respond by doing a better job if given an overview of the job's purposes. It is important to bear in mind that some *specific measures required by a landowner may affect price.*

In dealing with buyers, it is important for a landowner to keep in mind that harvesting wood products is a very difficult business. It requires a sizable investment, it is very dangerous, and it usually provides a low margin of return. It is subject to long layoffs because of weather and sudden downturns in log prices and demand. Logging is also very competitive, with many buyers vying for a limited resource. In order to make a profit, a logger must operate quickly without compromising the long-term productivity of the land, or the desires of a landowner for a job to be performed in a specific manner. A desirable buyer is one who pays a fair price, uses equipment carefully, and treats the land well and landowner fairly.

Road Systems

After locating a buyer, the owner's attention must turn to the design, location, and maintenance of a road system for extracting the trees to be cut. Ideally, road locations should be described in the management plan for a property, although on-the-ground considerations at the time of harvest often dictate some deviation from the plan. The seller or seller's agent should be actively involved in road planning because the roads and landings that might be most economical for the buyer are frequently not identical to those that would do the least damage to the site and that would provide long-term access for recreation or management. Under the usual arrangement, the buyer constructs roads which the owner or the forester has laid out and maintains them during and immediately after logging according to the owner's or forester's instructions.

Many woodland roads in New England are built with bulldozers, and many loggers own bulldozers and can install woods roads. If a buyer does not own the necessary equipment, there will be a number of local excavating contractors who do. Most excavating contractors charge by the hour for their work; the hourly rate normally includes the cost of the machine and the operator. Selection of a contractor should not be made solely on the basis of the hourly rate. The size and versatility of the equipment, as well as the operator's past experience and local reputation, should be considered. A large bulldozer with a power-angle blade may be able to do twice the work per hour, at only one and a half times the hourly rate, as a smaller machine. The operator's willingness to do the job according to the owner's or the forester's specifications is also important.

Because conventional skidding or forwarding equipment is very expensive to operate, and small machinery or animals have limited efficiency over long skidding distances, it is important for the buyer to be able to drive a truck as close to the harvesting site as possible. A *heavy-duty haul road* that accommodates modern, self-loading log trucks with three or four axles may be necessary. Such roads must be constructed on a solid base—frequently gravel—to support heavy use, and are usually expensive to build. Less costly *light-duty haul roads* will accommodate firefighting equipment in case of forest fires, and pickup trucks and farm tractors for the extraction of short-length firewood. The approval of the town highway superintendent may be necessary to bring an access road onto a town road. The town may require the installation of a culvert or a reduction in grade to minimize problems of water and mud buildup at the junction.

The present or future value of the trees to which a road gives access must be balanced against the costs of the road if the road's purpose is limited to the transportation of forest products. If a road is intended to improve access for recreational or other purposes as well, revenues from wood sales can be regarded as a subsidy for developing those uses. For tax purposes, it may be easiest to require the buyer to construct any needed roads as a condition of the sale. This expense reduces income from the sale, but eliminates some of the complexities in dealings with the IRS. The landowner or landowner's agent must supervise road construction to see that design standards are met.

Harvesting requires skidding routes over which trees are moved from where they are felled to a *landing,* a cleared area where logs are piled that is accessible to log trucks. A *skid road* is the main artery for moving several loads (or *hitches*) of logs; a *skid trail* is a secondary route by which one or a few trees are moved to the skid road (fig. 62). Like haul roads, skid roads and trails should be designed for long-term use, with both current and future operations in mind. A simple example of a woods road system is shown in figure 63.

A good landing is large enough to allow for unhitching, piling, bucking, and sorting logs as well as for maneuvering skidding equipment and log trucks. The exact size of a landing will depend on the available space, the type of equipment being used, and the complexity of the product mix; landings vary from small (40 feet by 100 feet) to several acres for whole-tree chipping operations. A landing should be on level, well-drained ground and well removed from any bodies of water.

Fig. 62. A skid road.
*Photo courtesy of
USDA Forest Service.*

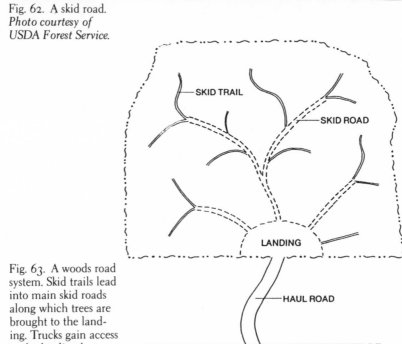

Fig. 63. A woods road
system. Skid trails lead
into main skid roads
along which trees are
brought to the land-
ing. Trucks gain access
to the landing by
means of a haul road.

Roads and their abuse, not the felling of trees, cause the most damage during logging. The importance of proper location, construction, and maintenance cannot be overemphasized because roads and landings are the most significant sources of erosion in any logging operation, and poorly developed roads are the most serious cause of water pollution in forestry.

Many laws governing logging operations in New England concern the control of water pollution. When leaf litter no longer protects soil from the impact of rain or the flowing of surface water, soil and humus particles are easily transported by water running off the land. In New England the fertility of the soil is in its surficial layers, and loss of topsoil has a significant and long-term impact on the quality of a site for growing vegetation, and thus on future timber values. In addition, if the soil that is carried away is eventually deposited as sediment in a stream, water can be rendered unsuitable for human consumption, fish habitat, and even industrial use. Unacceptable levels of sedimentation begin at only 5 parts per million (ppm) for domestic use and 70 ppm for fish. From the landowner's point of view, erosion also has financial consequences. A badly eroded road makes future forest management and firefighting operations difficult or impossible. Rebuilding an access road may be prohibitively expensive.

Table 20 Recommended Distances between Logging Roads and Bodies of Water

Slope of Land Between Roadsite and Water (percent)	Minimum Distance Between Roadsite and Water (feet)
0	25
10	45
20	65
30	85
40	105
50*	125
60	145
70	165

Source: R. F. Haussman and E. W. Pruett, *Permanent Logging Roads for Better Woodlot Management* (Washington, D.C.: U.S.D.A. Forest Service, State and Private Forestry, 1978), p. 2.
* Slopes steeper than 50 percent should be avoided.

Each New England state has a manual that specifies "best management practices" (BMPs) or "acceptable management practices" (AMPs) in the design, construction and maintenance of forest roads. The manual can be obtained from the state forestry agency through a county forester (see appendix 4).

Using topographical and stand maps, a road system should be first designed on paper, then evaluated, corrected, and marked out by flagging in the woods. Access roads are almost never surveyed, but are laid out by the landowner or forester and the contractor. Usually flagging is tied to trees or hung from branches to guide the equipment operator. The following eleven principles should guide road layout:

1. The road system must be kept to a minimum total mileage. A rough rule-of-thumb is that no more than 10 percent of the total area of the tract being harvested should be disturbed by roads. (Acreage of roads can be calculated by multiplying the width of the roadbed in feet by the total road length, and dividing by 43,560.) Adherence to this principle will minimize both the amount of land which will be incapable of producing timber for some time and the amount of damage to the remaining trees. A clear purpose should be definable for each road segment.

2. All roads should avoid slopes of 50 percent or more, as well as wetlands and ledges. Field-checking a route is especially important for avoiding wet and rocky places, since these may not be evident on topographical maps.

3. Roads should be set back from streams. The steeper a slope adjacent to a stream, the farther the road should be from the stream; see the recommended distances in table 20.

4. Stream crossings should be kept to an absolute minimum; should be located where banks are gentle, low, and solid; and should be at right angles to the stream bed (fig. 64). Crossings for haul roads will probably require culverts or bridges, as will skid roads where water quality for human or livestock use or fish habitat is to be maintained. When a crossing structure is necessary for these reasons, it can be a temporary culvert or bridge installed by the logger. If the landowner wants the logger to construct a permanent crossing, the labor involved will affect the price that the logger can offer for timber. In future sales, however, a pre-existing crossing may improve tree values. For crossing narrow streams, culverts are generally the most practical. However, a permanent, well-built bridge will probably outlast a culvert and have far less impact on the stream. When a crossing structure is necessary for a wide stream, bridges are sometimes the only practical choice. (See appendix 3 for state regulations regarding stream disturbance.)

5. Whenever possible, roads should follow contours of slopes. They can be tilted slightly outward to improve drainage, as illustrated in figure 65, but should be level on sharp curves where safety is a concern.

6. The ideal slope, or grade, of a road is between 3 and 6 percent. A 3 percent grade is enough to discourage the collection of standing water on road surfaces. On grades greater than 6 percent, water flowing down the road will have sufficient erosive effect to necessitate drainage devices. Grades up to 10 percent are acceptable for distances of 200 feet or less on haul roads; grades up to 20 percent are possible for distances of 100 feet on skid roads. For longer distances on such grades a system of drainage devices will be necessary.

7. Straight, sloping stretches of road should either be kept very short or drained with culverts placed at the intervals prescribed in table 21. It is frequently necessary on skid roads to balance control of the speed of drainage water with the straightness needed to minimize the damage to standing trees that may occur when a load is towed around corners.

8. Potential collection of water on the surface of a road should be prevented with low-cost water bars or broad-based dips; these are depressions in the road, usually made with a bulldozer, that collect water and divert it off the road surface. Water bars and dips are illustrated in figures 66 and 67. Prescriptions for their spacing are given in table 21.

9. Water that flows toward a road from an adjacent slope should be carried off in a ditch dug parallel to the road which is deep enough to handle the expected volume of water without spilling it onto the road (fig. 68). Culverts can be used to move ditch water under a road before it gains sufficient volume and speed to flood the road or to cause serious erosion. For haul roads, such culverts are usually made of corrugated steel, concrete, or plastic. On lower-class woodland roads, open-top or pole culverts may be sufficient (fig. 69). The size and number of relief culverts that are needed depend on the anticipated volume of water and the road grade (table 21).

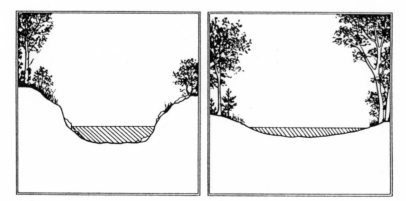

Fig. 64. Poor (*left*) and good (*right*) locations for fording streams with logging equipment.

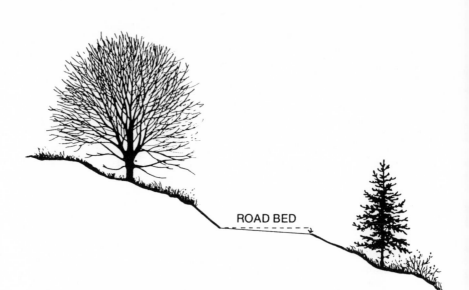

ROAD BED

Fig. 65. Outsloping roads. Haul and skid roads that follow hillside contours can be constructed so that they tilt slightly away from the slope to provide natural drainage of water that might accumulate on the road surface.

Table 21 Recommended Distances between Drainage Structures on Logging Roads

Road Grade (percent)	Feet		
	Recommended Distance Between Water Bars	Recommended Distance Between Culverts	Recommended Distance Between Dips
1	400	450	500
2	250	300	300
5	135	200	180
10	80	140	140*
15	60	130	—
20	45	120	—
25	40	65	—
30	35	60	—
40	30	50	—

Source: R. E. Hartung and J. M. Kress, *Woodlands of the Northeast, Erosion and Sediment Control Guides* (Broomall, Pa.: U.S.D.A. Soil Conservation Service and U.S. Forest Service, State and Private Forestry, 1977).
*Because of construction characteristics, broad-based dips should not be installed on roads with a grade greater than 10 percent.

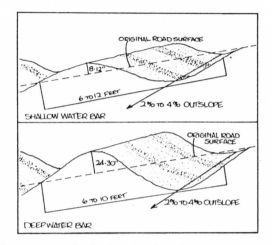

Fig. 66. Water bars. Water bars are mounds of soil angled downslope across a skid road or trail that divert water that might accumulate on the road surface. They may be either shallow or deep. Shallow water bars may be constructed prior to and during logging. Deep water bars are installed when logging is finished. *Courtesy North Carolina Division of Forest Resources.*

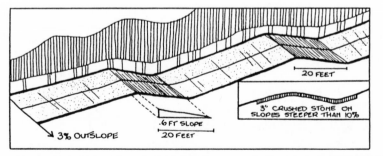

Fig. 67. Broad based dips. Broad based dips provide, if properly installed and maintained, an excellent means of diverting water from a haul and skid road surface. They cannot be used in roads with a grade exceeding 10 percent. Construction of broad based dips occurs during road construction, before logging begins. *Courtesy North Carolina Division of Forest Resources.*

Fig. 68. Ditches and ditch relief culverts. Ditches placed on the uphill side of a haul road will collect water running off the slope toward the road. Ditch relief culverts installed during road construction and placed at intervals along a haul road drain the ditches and move the water under the road.

10. Ditches should not terminate where they will feed water directly into streams; ditches should divert water into the woods, or into a siltation basin.

11. Log landings should be located (A) as close as possible to the stands being harvested, in order to minimize skidding distances; (B) on slopes of less than 5 percent; (C) at least 200 feet from ponds, lakes, streams, or marshes.

Harvesting Trees

In all but roadside sales the buyer is responsible for harvesting the designated standing trees, a process that has four steps: *felling, limbing, bucking* and *transportation.*

Felling means cutting down the trees. The modern options for felling equipment are the manually operated chain saw, the shear, and the machine-mounted chain saw. On woodlands where a variety of species and sizes of trees are being harvested, the manual chain saw is most common. A skilled operator, or "chopper," can fell a tree with a great deal of control. Accurate directional felling is a key to minimizing damage during removal of harvested trees. Shears and machine-mounted chain saws are hydraulically operated attachments to any one of a number of pieces of equipment (fig. 70). Because they are remotely operated from an enclosed cab, automated felling devices are safer than manually operated chain saws. Automated felling is most common in softwood stands on flat and moderate terrain.

Limbing involves cutting the limbs off a tree and lopping the top off at the point where the tree stem becomes too small in diameter to be merchantable for the products being harvested. Most limbing in New England is done with a chain saw at the spot where the tree falls, usually by the chopper who fells it.

Bucking refers to cutting a tree into shorter lengths—*sawlogs, bolts,* or *sticks. Sawlogs,* which in New England are between 8 and 16 feet long, are made into lumber. Softwood trees are bucked into sawlog lengths of 8, 10, 12, 14, or 16 feet; hardwoods may be bucked at any 1-foot interval between 8 and 16 feet, with an extra 2 to 6 inches added for a so-called "trim allowance." A *bolt* is a short log, usually between 4 and 8 feet long; bolts are used to make short lumber pieces for pallets or furniture. Lengths of pulpwood, used in papermaking, vary between 4 and 8 feet but it is common for pulpwood to be delivered in longer lengths and bucked at the receiving location. Logs used for railroad ties or veneer may be odd lengths. Local mill specifications will determine appropriate log lengths (see table 3). Most bucking, like limbing, is done with a chain saw, although there are more mechanized alternatives available.

Integrated harvesting machines that fell, limb and buck are available and are becoming more common. These sophisticated pieces of equipment are best suited to fairly uniform softwood stands on moderate terrain. The operator controls all functions from an enclosed cab.

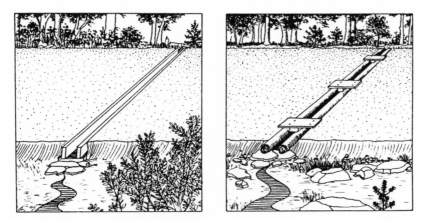

Fig. 69. Open-topped and pole culverts. These culverts are relatively inexpensive devices for removing water from skid trails, skid roads, and sometimes, haul roads, and for draining small seeps across roads. They are usually installed as needed during logging, or placed in problem spots after logging is finished.

Fig. 70. A hydraulic shear attached to a feller-buncher. *Photo courtesy of Morbark Industries.*

Transportation falls into two classes—primary and secondary. Primary transportation moves a tree from the place where it was felled to the landing. Secondary transportation moves wood from the landing to the mill on a truck.

Primary transportation is accomplished by skidding trees or by carrying them clear of the ground (forwarding). The most common skidding equipment in New England is the rubber-tired cable skidder, which is a large, four-wheel drive machine with tires specially designed for the woods. A winch pulls felled trees to the machine, then suspends one end of the trees above the ground to facilitate dragging (fig. 71). On a grapple skidder, the winch and cable is replaced by a hydraulically operated attachment that grabs and picks up felled trees (fig. 72). A grapple skidder can pick up and drop trees much faster than a cable skidder, but does not have the ability to pull a tree in by cable from 75 feet away. Crawler tractors, which are really small bulldozers outfitted with a winch and cab, were popular for skidding before the development of the cable skidder. Farm tractors can also be equipped with winches or homemade rigging for skidding logs, and such an arrangement is common among farmers and others who do their own woodswork and who use tractors for other tasks (fig. 73).

Like skidding, forwarding is a function that can be performed by any one of several machines or animals. The machine commonly called a "forwarder" is a large, four-wheel-drive vehicle equipped with bunks and a

Fig. 71. A rubber-tired cable skidder. *Photo courtesy of John Deere.*

Fig. 72. A grapple skidder. *Photo courtesy of John Deere.*

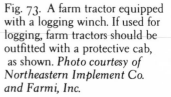

Fig. 73. A farm tractor equipped with a logging winch. If used for logging, farm tractors should be outfitted with a protective cab, as shown. *Photo courtesy of Northeastern Implement Co. and Farmi, Inc.*

grapple that can pick up logs in the woods or at a landing (fig. 74). A loaded forwarder is more maneuverable than a skidder with a hitch of trees behind it. However, a forwarder is less stable than a skidder, especially when loaded, so it is confined to gentler terrain. Crawler or farm tractors equipped with a boom and scoot or a wagon can be used to forward logs. A boom is a device consisting of a pole extended from the rear of a crawler or farm tractor along which runs a cable with tongs. A scoot, which resembles a large, low sled is towed behind. Logs are dragged up onto the scoot with the cable and tongs. Booms and scoots are increasingly rare in commercial logging operations. Draft animals can also be used to forward short-length wood.

The type of equipment used in harvesting determines in part the impact of a logging operation on a forest. What equipment a particular logger uses should be considered. A decision usually involves a trade-off between high daily productivity and impact on the site. *However, the extent of logging damage is determined more by the way in which a machine or animal is used than by the type of equipment.* Significant logging damage results from careless or unskilled operation of *any* skidding or forwarding machinery, from poor handling of draft animals, or from poor road layout and careless felling. In general, the larger the equipment, the greater its potential impact on the land and residual trees.

Skidders and forwarders move faster than other equipment, and have the capacity for high daily productivity. However, their tires, which are designed to handle the roughest terrain, have a tendency to tear up woods roads. On steep slopes, wet areas, and fragile soils, their use can result in erosion, soil compaction and rutting; where skidders are used, such damage may be intensified by the large volume of wood being dragged behind. The length and breadth of skidders and forwarders inevitably results in bark being scraped from some remaining trees.

Crawlers are slow, and therefore uneconomical if the skidding distance is much greater than 800 feet. However, they are very maneuverable in the woods, especially when used with a boom and scoot. In addition, their weight distribution and lower center of gravity generally give crawlers better stability on steep inclines. Farm tractors are very dangerous in the woods: they do not have an enclosed cab, and are much more unstable on slopes than skidders or crawlers. With a hitch of logs pulling down the rear end, there is also danger that a farm tractor will flip over backward. Horses,

Fig. 74. A forwarder. *Photo courtesy of Tree Farmer, Inc.*

Fig. 75. A whole-tree chipper in operation. *Photo Courtesy of Morbark Industries.*

oxen, and ponies—the oldest sources of power for skidding—are slow, can move only very small payloads on each turn, are of limited use in winter, and can operate economically only over short distances. However, at the hands of a skilled teamster they can leave the woods minimally disturbed. In addition, they are quiet, in contrast to motorized machinery.

Harvesting systems that have a minimal impact on the land often require better roads than other systems. For instance, on roads used by draft animals, it is important that stumps be cut down to ground level and that all brush be cleared from the skid road. Where crawlers are working, it is desirable to remove large rocks from roads to reduce wear on the tracks and to make the ride less rough. Farm tractors need good, smooth roads to minimize the danger of tipping over. Draft animals and farm tractors are severely limited where roads ascend or descend extreme slopes.

Harvesting wood for chips to fuel biomass energy facilities has become common in northern New England. "Energy wood" most often comes from integrated harvest operations that also include sawlogs and pulpwood. Trees may be felled manually or by feller-bunchers, which are wheeled or tracked vehicles equipped with a hydraulic shear. The feller-buncher can snip a tree off at ground level and place it into a pile, or bunch (fig. 70). Commonly a grapple skidder works with a feller-buncher to skid the entire bunch to a landing where a whole-tree chipper converts the bunch, branches and all, to matchbook-size chips (fig. 75). The chips are blown directly into vans and trucked to a wood-burning facility. The total fiber yield of "whole-tree harvesting" may be twice that of conventional harvesting systems.

Moving a whole-tree chipper to a landing site is an expensive fixed cost. Sorting the various products that come from most logging jobs, as well as chipping and blowing whole-tree chips into a trailer requires a good deal of landing space. As a result, operations that involve chipping are more viable on larger acreages. Several hundred tons of chips are needed to attract an operator. As a rule, harvesting needs to occur on at least 30 acres to make whole-tree chipping viable. If some of the acreage is being clearcut, a lower minimum will apply.

When tops of trees are extracted from the woods in a whole-tree operation, there is increased likelihood of scraping bark off remaining trees because of the width of the "hitch." Extra care must be taken to avoid excessive damage.

Nutrient depletion is an important issue in whole-tree harvesting. Conventional harvests, which remove only a portion of the main tree stem and only at infrequent intervals, do not result in significant nutrient removals from the site. Most of the biomass is left behind and the concentration of nutrients is much higher in branches, twigs and leaves than it is in the main stem. When the entire above-ground portion of the tree is removed, the flow of nutrients off the site is much greater. As a general rule, to avoid depleting the fertility of a site, whole-tree clearcuts in New England should not be more frequent than every 50 years. Whole-tree harvesting is advisable for intermediate treatments by a skilled operator (fig. 76).

Contracts for the Sale of Trees

No landowner should sell standing trees without having a good contract with the buyer. A review of any proposed contract by an attorney may be helpful. A contract clearly assigns responsibilities and liabilities to buyer and seller, outlines the seller's standards for the logging job, and provides criteria for evaluating the job.

Fig. 76. A site immediately after a thinning accomplished with whole-tree harvesting equipment. *Photo by John Herrington.*

Contract clauses can be composed to cover the particulars of any sale, but all contracts should cover a nucleus of issues. Sample wording for the important conditions of a sale contract is given below, followed by annotations where appropriate. The main subjects of clauses are indicated by italicized headings. Note that the sample clauses assume that a consulting forester has been employed to act on the owner's behalf. When this is not the case in an actual contract, the word "forester" and the phrase "agent of the seller" can be replaced with "the seller." This sample contract is written for a lump-sum sale; unit-price contract wording and annotations are provided where appropriate.

DATE AND PARTY IDENTIFICATION

Agreement made this ___(date)___ of ___(month)___ , ___(year)___ , between ___(landowner's name)___ of ___(landowner's town or residence)___ , hereinafter called the seller, and ___(buyer's name)___ of ___(buyer's town and state)___ , hereinafter called the buyer.

As in any type of contract, the sales contract is dated and the parties are identified.

Sale Volume

(a) The seller agrees to sell to the buyer, upon the terms hereinafter stated, all trees marked or designated by the agent of the seller,   (name of forester)   , the net volume of said trees estimated to be   (number)   thousand board feet of sawlogs,   (number)   cords of pulpwood,   (number)   cords of fuelwood, and   (number)   cords of wood to be chipped, more or less, and located on a tract of land belonging to the seller in the town of   (town and state)   . Seller does/does not retain ownership of tops.

In a lump-sum sale, the amount and location of what is being sold is described. In a unit-priced sale, estimated volume may or may not be specified. The units of volume used must be legal for the state in which the land is located. Ownership of hardwood tops should be clarified. Before fuelwood became valuable, the wood in tops of harvested trees was worthless. Recently, follow-up fuelwood operations in which tops are sold to a different buyer have become common. To avoid any confusion, a contract should specify whether the tops are included in the sale. In the case of full-tree logging, tops are always included, since the entire tree is usually sold and chipped.

(b) All trees, and only trees designated by   (color)   paint at about 4.5 feet above ground and at stump height, shall be cut. Trees marked with a slash are designated as sawlogs; trees marked with a spot as pulpwood or fuelwood; trees marked with an "X" are cull trees and shall be girdled with a chain saw, or removed.

The forester's or owner's marking system is described. The stipulation that cull trees be treated by the buyer is not an unusual one, but it is an example of a contract condition which requires extra labor on the part of the buyer, and could therefore affect the stumpage price offered. Occasionally where all low-value trees in a given area are to be cut, the boundaries of the area will be delineated with paint and the volume of trees within the bounded area determined by a sampling procedure. In this case, not every tree will be marked with paint.

(c) For the purposes of determining sawlog volume, sawtimber was scaled by the   (log rule name)   log rule to a top inside-bark diameter of   (number)   inches. Pulpwood and fuelwood were scaled by the   (table name)   table to a top inside bark diameter of   (number)   inches.

Both the choice of the log rule used to determine wood volumes, and the point on the tree stems past which the stems are considered too small to use and measure, are important factors in substantiating the calculation of the sale volume. The log rule is named in the contract because volume estimates for the same tree will vary among rules. The top inside-bark diameter is at the point on a tree stem that would be the small end of the uppermost log in the tree, and it is dictated by local sawmill standards. It is usually a diameter of 6 to 8 inches for sawtimber, and 4 inches for fuelwood and pulpwood. Because chips are usually made from whole trees, no upper stem diameter is stated. The point where the tree tapers to the minimum top diameter is estimated by eye from the ground, and the estimate is adjusted slightly downward to deduct for the bark. In a unit-priced sale, the volume involved in the sale may be estimated in the contract; however, the final volume on which the buyer pays is determined at the landing or the mill.

(d) Defect in sawtimber was deducted in the field at the time of marking. Defects considered were excessive crook, sweep, rots, seams, and large limbs.

This clause makes it clear to the buyer that the sale volume quoted is a net amount; that is, the volume of obviously defective wood in each tree is not included in the volume estimate. In a unit-priced sale, deduction for defects will occur after a tree is removed from the woods. A danger in a unit-price sale is that a logger will leave a partly defective tree in the woods because there is no real incentive to recover the sound volume in the tree. Crook is an abrupt angle in a relatively short section of a log. Sweep is the overall bow in a log, from end to end.

(e) The merchantable volume is not guaranteed, but is the result of individual tree measurement and is final for this sale.

It is important to remember that in a lump-sum sale, the seller is selling designated trees for a specified amount of money. The volume figure in the contract is an estimate only. Different loggers may end up with varying yields because of differences in bucking practices or utilization. Also, unseen defects may affect the final yield. This contract clause does not apply to a unit-priced sale.

Declaration of Ownership
The seller warrants that:
1. He/she has the full legal right to sell the trees described in this agreement.
2. There are no mortgages or encumbrances affecting the sale of the trees covered by this agreement, except as may be listed on a separate sheet attached to this agreement.

3. The title of the marked and designated trees is guaranteed to the buyer, and the seller will do nothing during the term of this agreement to interfere with or jeopardize the rights of the buyer to said marked or designated trees.
4. Property lines necessary for this sale are clearly and correctly marked.

This clause assures the buyer that the landowner in fact owns the wood being sold, and/or has the legal right to sell it. It guarantees that the ownership of the trees will be transferred to the buyer with no other conditions than those in the contract. An important function of this clause is to protect the buyer from liability for timber trespass, in which trees are harvested from an abutter's land because inaccurate boundary information was provided by the seller.

Duration of the Contract
The seller hereby grants permission to the buyer to enter the above designated tract to cut and remove the marked and designated trees. The term of this agreement shall be for __(number)__ months from the date first written, and all rights of the buyer shall terminate at the completion of this term. At the termination of the contract, title to any remaining trees or logs remaining on the property reverts to the seller. If, because of extenuating circumstances, the buyer cannot complete this contract, the buyer shall apply in writing for an extension at least one month before the expiration of the contract. Should an extension be granted, it may be subject to a reappraisal of the contract prices.

Setting a period of time during which the buyer is allowed access to the land prevents delays in the start-up and completion of logging, especially when log markets become unfavorable for the buyer. Limiting the duration of the logging operation also helps to minimize disturbance to the site and the time needed for its recovery.

Agreed Price
The buyer agrees to pay the seller for trees cut under the terms of this contract a lump sum of __(dollars)__ . Payment shall be made in __(number)__ installments, each in the amount of __(dollars)__ . The first installment shall be due upon the signing of this contract. The second installment shall be due at the beginning of cutting. Additional installments shall be due __(interval between payments is defined)__ . All trees shall be paid for before the contract expires or is terminated.

The above is typical wording for a lump-sum sale. Lump-sum contracts often specify that payments will be made when a certain volume has been cut, or within a certain time period, whichever comes first. It is critical in a lump-sum sale that all trees are paid for before they are cut. When trees are sold by unit pricing, the clause might read as follows:

The buyer agrees to pay the seller for the above estimated, marked, or designated trees at the bid rate of
_____ (Species) _____ (Product) _____ (Price per cord/MBF) _____
Payments for trees cut, during or within a calendar month, shall be received on
or before the tenth day of the calendar month following said cutting.

Weekly, biweekly, and monthly payment schedules are common in unit-priced sales. If, in a unit-priced sale, payment is made on the basis of the mill tally, a clause such as the following must be added:

Payments by the buyer must be accompanied by copies of all scale slips for the
trees being paid for.

When a load of logs arrives at a mill, each log is measured and the volume is recorded on a scale slip. The landowner should receive copies of all slips.

Interest on Late Payments
Interest will be charged on any late payments at the rate of __(percent)__ ,
compounded and calculated __(frequency)__ .

While the contract may allow cancellation of the contract for payments not made on time, this clause allows the buyer some leeway, and protects the seller without forcing cancellation.

Performance Bond/Deposit
The buyer agrees to deposit with __(forester's name)__ , as agent for the
seller, the sum of __(dollars)__ upon the signing of this agreement by the
buyer. This deposit shall not be used as a credit for trees cut, but, on completion of the sale and full compliance with all the terms of this agreement by the
buyer, the sum shall be returned to the purchaser. The deposit shall be in the
form of a certified check or bank draft.

The requirement of a performance bond is always advisable. The amount is often set as 10 percent of the estimated total value of the sale. In the case of a deposit, the forester or a third party holds the deposit in escrow, return-ing it to the buyer only after the logging has been complete and a final in-spection by the forester or landowner assures that all terms of the contract have been fulfilled. In order to assure that the buyer's commitment to clean-up, site restoration, and erosion control work will be carried out as agreed, a deposit is not treated as a final payment for the timber. A bond is purchased from an insurance company. The buyer must pay a small per-centage of the face value of the bond, and is then insured for the full amount. There are two principal problems with bonding. First, a small or new logger may have difficulty getting an insurance company to sell him a bond at all. Second, if it should become necessary for the forester or landowner to use the bond money to correct job deficiencies or to pay for excessive damage, he or she must sometimes prove to the insurance company's satisfaction that a contract violation did indeed occur. If the insurance company chooses to fight the forester's or owner's contention, a protracted discussion may result. In the case of a performance deposit, on the other hand, the forester or third party holds the full amount of money and has discretion over its use.

Agent Authorization

  (Forester's name)   is hereby appointed as agent for the seller for the following authorized purposes: to inspect the progress of any cutting of trees and work done under this contract; to receive and disburse all monies associated with this sale; to halt the job if a contract violation occurs; to act on behalf of the seller to ensure the successful completion of the contract.

When a forester has been retained by the buyer as agent, this clause defines the agent's role and authority. The agent is given both specific and intentionally general powers and responsibilities. Harvesting operations run most smoothly when the forester's expectations are made clear and when the logger understands them.

Default

In the event the buyer defaults in payments that are due or fails to fulfill any of the terms of this agreement, the seller or seller's agent   (forester's name)   upon written notice to the buyer shall have the right to stop further cutting by the buyer; and immediately thereupon title to cut, marked, or designated trees shall revert to the seller; and the seller shall be entitled to retain the whole of the performance deposit as part of damages for breach of this contract; and the agreement shall be automatically terminated.

This clause spells out the seller's right to terminate the agreement for breach of contract. Retention of performance money by the seller is necessary because of the costs associated with a partly completed logging job. In addition to clean-up measures that may be required, the remaining uncut trees may be difficult to resell, and will usually bring a lower price than in the original sale.

Authorized Cutting of Undesignated Trees

All trees not designated for cutting are reserved from cutting, except undesignated trees felled in the construction of landings or skid roads or inadvertently damaged, as approved by the forester. Such trees may, in accordance with agreement between buyer and seller, be cut and removed and paid for at the rates indicated below:

| (Species) | (Product) | (Price per MBF/Cord) |

In virtually every logging job, some trees will be damaged by even the most skillful operator. In addition, some unmarked trees may be in the way of operations. In either case, the owner or forester should be informed and should decide if compensation should be made for those trees which have commercial value.

Unauthorized Cutting of Undesignated Trees

Undesignated trees that are severed or unnecessarily damaged during logging shall be paid for at __(number)__ times the rates set forth in this contract. The buyer is not liable for damage which is customary in the harvesting of timber/fuelwood/pulpwood.

The penalty for the unapproved cutting or unnecessary damage of an unmarked tree is usually three to ten times the sale value of the tree, as calculated by the prices set forth in the contract. The volume of wood of a missing tree can be roughly estimated as equal to that of neighboring trees of the same species that have similar diameters at stump height. Foresters use mathematical equations for making such estimates.

A few merchantable trees will be damaged during logging. However, where the amount of damage is considered excessive by the forester or landowner, compensation may be required. Assessing logging damage is discussed in the evaluation section below.

Cutting Procedure

All marked or designated trees shall be cut and removed. No re-entry shall be permitted without the approval of the forester. When operating in any one area, all marked or designated trees shall be removed. The buyer will make every effort to protect unmarked trees and regeneration during harvesting.

The clauses on authorized and unauthorized cutting preceding this one require that *only* designated trees are removed; this clause on procedure ensures that *all* marked trees are taken. This clause is important, because the choice and number of trees marked was largely determined by the light, moisture, and growing space requirements of the remaining trees, or the desired regeneration. Designated trees which are left may limit the success of management of the remaining stand.

This clause also requires the logger to cut all marked trees in one area before moving on to the next area. Once an area has started to recover from logging disturbances, and especially when desirable regeneration has become established, much damage will be done if logging equipment is reintroduced. In the case of an installment sale, this clause prevents the buyer from cutting only high-value trees and then defaulting on the remainder of the contract.

Utilization Standards
The buyer shall utilize each tree to its lowest merchantable diameter for the product(s) harvested from each tree. Softwood sawtimber will be utilized to a 6" top, hardwood sawtimber to an 8" top, fuelwood and pulpwood to a 4" top.

The utilization standards clause is meant to ensure that the logger minimizes the amount of slash left in the woods. Slash can be visually offensive and can increase the fire hazard, especially in softwoods. In a unit-priced sale, this clause is critical. Since the buyer is only paying for what leaves the woods, he or she must be pushed to use material to the lower limit of merchantability. In a lump-sum sale, the buyer is paying for the volume, regardless of whether it is utilized. Crookedness, rot, and other defects will often dictate that material which would otherwise be merchantable be left in the woods.

Notification of Cutting
The buyer shall notify the forester at least __(number)__ days in advance of the beginning of the logging operations.

Notification of cutting is important if a forester is involved, or if the landowner is not likely to otherwise know if logging has begun. Early inspections of logging jobs should be frequent to avert irreparable difficulties and to maintain good communications. This clause also allows the forester to advise the logger when start-up should be delayed because of deep snow, mud conditions, or the danger of fire.

Equipment Restrictions
The seller reserves the right to restrict the size and type of log-skidding equipment, and the manner in which the equipment is operated, if the seller or the seller's agent determines that excessive or unreasonable damage is being caused by the size, type, or manner of operation of the equipment.

Although the type of equipment used is sometimes a factor in selecting a logger, this clause gives the landowner recourse if unexpected damage from equipment is occurring during an operation. It also serves to prevent the logger from using substitute equipment that is undesirable to the landowner. This second purpose might also be met by a clause that names the types of equipment which shall, may, or shall not be used for bunching, skidding, or forwarding trees. The degree of damage from skidding can often be controlled by limiting the length of materials that may be skidded;

lengths of 25 and 33 feet are common when it is necessary to impose such a restriction. Limiting the length of skidded material dramatically increases the costs of harvesting. If such a restriction is imposed for the entire operation, it should be clearly stated before price is agreed upon. Limiting the length of skidded material is a radical step that will greatly affect the economics of the operation for the buyer, and thus once the logging job is underway, its use as a corrective measure should only be imposed as a last resort.

Seasonal Restrictions
The following restrictions apply to the timing of harvesting operations: _____
_____
_____

This clause may be used either to prohibit or mandate operations at key times for wildlife protection purposes, protection of regeneration, or to create conditions suitable for regeneration. Prohibition of logging during nesting season for specific species of birds would be the most common use of this clause for a wildlife purpose. If the aim were to protect very young seedlings, an owner might require that logging take place when there is a snow cover. If creation of a mineral soil seedbed to promote regeneration of a particular species is the goal, this clause might require that logging take place in a season other than winter.

These types of restrictions may affect the price a landowner can expect to receive for timber, since a logger may have to cease operations, move off the job and move back weeks or months later. Ceasing and restarting operations can represent a significant cost.

Roads
Logging roads shall be located and constructed under the direction of the forester. The buyer shall leave all roads in proper repair, as approved by the forester, at the completion of logging.

Although brief, this is one of the most important contract clauses, since the proper location, construction, and maintenance of roads are the three most crucial factors in minimizing the environmental impacts of logging. The clause makes clear that road design and maintenance will be under the direction of the seller or the seller's agent.

Stream Crossings
The buyer shall not ford any brook, stream, or other watercourse except where designated by the forester, and shall minimize the disturbance to the streambed insofar as may be practicable.

If it is not possible to locate crossings where stream banks are low and stable and the streambed is firm, or if the maintenance of streamwater quality is especially critical, this clause might be amended as follows:

The buyer shall not cross any brook, stream, or other watercourse except where designated by the forester. As directed by the forester, the buyer shall construct temporary crossing structures, approved by the forester, which minimize the disturbance to the streambed to the extent practical. The forester will stipulate whether, upon termination of logging activities, the structures are to be removed, retained, or replaced.

Buffer Strips
The buyer shall not remove, fell, or otherwise disturb vegetation within __(number)__ feet of a pond, lake, marsh, or stream, except as designated by *the forester.*

Vegetation along and around standing and running water maintains water quality by filtering sediment and nutrients from runoff, by stabilizing banks, and by keeping water temperatures cool, thus maintaining aquatic habitat. As a rule, where partial cutting is occurring within a watershed of 20 square miles or less, very few trees should be cut and no roads should be built within 25 feet of a watercourse or body of water. The buffer strip should be at least 50 feet wide on larger watersheds. The Soil Conservation Service maintains information about specific watersheds, including their acreages (see appendix 4). If harvesting is done by clearcutting, only light cutting should occur within 100 feet of water. However, state and local regulations should be consulted, since they vary widely.

Spills
The buyer shall have available on the site oil sorbent pads. Any spills shall be cleaned up immediately. Machine fluids shall be changed only at the landing while the machine is parked on a sorbent pad. Used fluids shall be collected and disposed of properly.

Logging equipment uses several types of petroleum products. The most common spills occur when hydraulic lines burst. Requiring the use of sorbent pads minimizes the risk of pollution to soil and water from operation of logging equipment. Loggers may be required to park machines on sorbent pads at night and during shutdowns.

Habitat Protection
The buyer agrees to operate equipment in such a manner so as to minimize disturbance to the following key wildlife habitat areas: winter deeryards; wetlands; vernal pools; and areas critical to rare and endangered species.

Especially in northern New England, closed softwood stands can be important to survival of deer. Vernal pools are isolated, seasonally flooded depressions critical to the breeding and survival of a number of amphibian species. Although other clauses in a contract deal with proper operation of equipment, it is useful to deal explicitly with the treatment of important wildlife habitats. Some states have laws governing the conduct of timber harvesting in or near these habitats.

Stumps

The buyer shall leave stumps no higher than the top of the root swell, except where there is metal in the wood, and stumps shall be cut so as to leave the forester's paint mark plainly visible.

Root swell occurs where the aboveground part of a tree joins the root system. It usually terminates within a foot of the ground. Stumps cut low maximize the utilization of the tree and greatly improve the appearance of the site. Metal in the wood includes barbed wire, nails, maple sap spigots, or other metal imbedded in trees that present a threat to loggers, sawyers, chain saws, and sawmill equipment. It is important that stumps be cut so that their paint marks remain to provide the owner or forester with assurance that the removed tree was intended for harvest.

Slash

No slash shall remain within __(number)__ feet of the rights-of-way of public roads, adjoining property lines, and railroad tracks. Slash shall be removed completely from roads, trails, and streams. All slash shall be lopped to within __(number)__ feet of the ground. No organic debris shall be used in the fill of any road or landing.

Provisions for the disposal of slash should be included in any contract. Most New England states have fire protection laws regulating the treatment of slash within a certain distance of public roads, property lines, and railroads; they also have water pollution control measures to regulate the treatment of slash in or near streams (see appendix 3). Although a contract should have a clause requiring the logger to abide by all pertinent laws, explicit wording is helpful and allows for stricter constraints than those prescribed by law.

For reasons of aesthetics, access, and perhaps forest fire control, it is advisable to stipulate that slash also be removed from roads and trails on the land. Slash that is lopped, or cut into short sections, will lie closer to the ground, be less visible, and be quicker to decay. However, lopping requires additional labor by the logger and may affect the stumpage price. In areas where aesthetics are not an important consideration and fire danger is low, lopping is unnecessary.

Slash that has been buried in roadbeds may eventually result in soil instability, since the slash will deteriorate in the subsoil. Where a road or a landing is known to be temporary and will not be used in the future, the burial of organic material may be acceptable.

Broken and Uprooted Trees

The buyer shall either remove or fell to the ground all trees which are broken, damaged, or leaning as a result of the buyer's activities in the building of roads and the cutting of designated trees. In the construction of all roads and landings, all trees more than three (3) inches in diameter shall be severed at the stump level, and not pushed over by logging equipment.

These restrictions are largely to protect aesthetic values, although leaning trees also present safety hazards, and the uprooting of trees with logging equipment may result in soil erosion. Determining when damage to a tree is sufficient to warrant its removal is discussed in the evaluation section below.

Fire Safety

The buyer agrees to observe all state fire laws, and to use due precautions to prevent forest fires, and to initiate immediately suppression activity on any fires which may occur on or adjacent to the sale area, using all necessary labor and equipment at the buyer's disposal. The buyer shall be liable for any claims arising from forest fires attributable to the buyer's operation while the operation is in progress. Skidding equipment and trucks shall be equipped with fire extinguishers.

New England state fire laws concerning timber harvesting are described in appendix 3.

Clean-up

All buildings and landings erected by the buyer during the operation shall be removed; all camping, lunching, and service areas shall be cleaned up, all within the time limit of this agreement. At log landing and loading areas, all forms of waste, including unmerchantable logs or portions of logs, shall either be trucked away or buried. These areas shall then be smoothed, leveled, and seeded where necessary as specified by the seller or seller's agent.

Seeding of landings, and sometimes of logging roads, is required of loggers with increasing frequency. "Conservation Mix," a commercially available combination of clover, grass, and fescue seed, is commonly used.

Suspension of Operations
Without penalty to the buyer, the seller's agent may suspend removal operations if the forester determines that unreasonable damage to access roads, skid roads, and logging roads is resulting from the use of these roads during periods of excessive ground wetness.

Contracts for winter logging jobs frequently call for the suspension of operations for some or all of the period from March 1st to June 1st. If logging is suspended for long periods, it is common to extend the contract correspondingly.

Risk of Loss
The risk of loss of all trees which are marked or designated in accordance with this agreement shall be borne by the buyer.

This provision would be important in lump-sum sales when there was loss of trees as a result of forest fire. In lump-sum sales, ownership of the marked trees changes hands when the contract is signed, even if only partial payment is made at the time of signing. Thus, the risk of loss rests with the buyer, who is the legal owner of the standing trees. In a unit-price sale, on the other land, ownership is transferred at the time the trees are measured (after they are cut). The risk of loss then rests with the seller until the trees are cut.

Disease or Insect Epidemic
In the event of a serious disease or insect epidemic, and only with the written recommendation of the state forestry agency, the seller may suspend the operation. The seller will return to the buyer any money paid for timber which cannot be harvested.

This clause is intended for extreme cases only. Logging during a severe insect attack, or during a disease epidemic, adds an additional source of stress to a stand. In addition, mortality resulting from an epidemic may dictate subsequent alternation of the silvicultural approach taken with the stand. It will usually be possible to avoid invoking this clause because most disease and insect epidemics can be predicted.

Liability of the Seller
The buyer shall hold the seller harmless and indemnified from and against any claims for any injuries or damages incurred by the buyer, its employees, or its associates, or by any third persons resulting in any way from the buyer's operations under this agreement. The buyer agrees to abide by all applicable federal, state, and local laws and bylaws.

Most laws and regulations pertaining to harvesting operations are state, rather than federal or local (see appendix 3). Most deal with filing requirements, fire prevention, and the protection of wetlands or water quality. An increasing number of towns, however, have local bylaws that attempt to regulate various aspects of harvesting. It is advisable to inquire about such regulations before beginning logging operations.

Buyer's Insurance
The buyer agrees to maintain public liability insurance and worker's compensation insurance in connection with its operation, and shall, upon request by the seller or the agent of the seller, file certificates of this coverage with   (forester's name)   , as agent of the seller.

Worker's compensation insurance is now available in some states to owners of sole proprietorships or to partnerships, which are the forms of organization used by many independent loggers. All employees can and should be covered.

Assignment of Work
The seller reserves the right to approve all subcontractors selected by the buyer. The buyer agrees to enforce upon his subcontractors all applicable provisions of this agreement.

Many buyers, especially sawmills, contract with independent loggers to do the actual harvesting of the purchased stumpage. It is important that the seller or seller's agent be allowed to reject loggers selected by the buyer if the logger is unable or unwilling to do the job as required. Also, the final responsibility for execution of the contract must be clearly assigned to the buyer.

Extent of the Contract
This agreement shall be binding upon the heirs, executors, administrators, successors, and assigns of the parties hereto.

The buyer, the seller, and, if desired, a witness for each, sign the contract. The forester employed by the seller does not sign the contract.

These sample clauses cover the most common and the most important elements of a sale contract. A contract can deal with any other general or specific concerns of either the buyer or the seller. As is true of any contract, however, even the most detailed written agreement might be breached on the basis of technicalities, and although a contract is essential, it is no substitute for good communication between the buyer and the seller.

Overseeing and Evaluating a Logging Job

Because an important element of good communication is its frequency, a logging job should be visited a few times over the first week of operations, and weekly thereafter. Inspections are made by the forester, if one is involved, otherwise by the landowner. At that interval, inspections can prevent undesirable situations from becoming serious or irreparable. It is easiest to establish a good working relationship between the buyer and the seller early in the course of a job.

Using the written contract with the buyer as a checklist for each weekly inspection, the landowner or forester should make sure that the following conditions are being met:

1. Haul and skid roads are draining properly and are in good condition. Water should not be running in any roadbed. The most common cause of erosion and stream sedimentation during logging is the use of a skid road or a trail under excessively wet conditions; operations should be suspended in wet weather. Culverts, ditches, or other drainage structures should be maintained in functional condition by the logger during the operation.

2. Stream water is not muddy.

3. Stream banks at crossings remain stable, and logging equipment is never operated in the streambed, except at designated crossings.

4. Care is being taken to minimize damage to the residual stand, and improper felling techniques are corrected. A judgment must be made whether too many residual trees are being damaged, and whether the owner should be compensated for the lost value according to the terms of the contract.

5. Critical elements of wildlife habitat are being safeguarded. Logging damage to wildlife habitat includes water pollution from soil particles, organic matter, and toxic substances, to which fish are very sensitive. Stream bank vegetation should not be disturbed. Other wildlife habitat elements commonly threatened by logging are wild fruit trees, den trees, snags, and softwood stands which are known to provide winter cover for deer. Harvesting operations should be checked regularly to make sure the operator has understood that these elements are to be safeguarded.

A final, follow-up inspection of a logging job is especially important. It should be undertaken immediately after the woodswork has been completed, and (ideally) before the operator has removed the logging equipment. The site should be very thoroughly checked to see that all contract provisions are satisfied. Particular attention should be paid to three conditions:

1. Roads and landings have been properly retired. All antierosion measures should have been installed to assure long-term soil stability. This may have re-

quired the replacement of temporary devices used during the logging operation with relatively permanent structures. Rutting caused by the wheels of machinery and the dragging of logs should have been smoothed to prevent the channeling of water down roadbeds.

2. No slash remains in any stream, lake, or pond, or near any property boundary. Slash has been removed from roads and landings, and has in general been treated according to the terms of the contract. Debris in a stream is likely to result in bank erosion, and will add excessive organic matter to the water as it decomposes.

3. Severely damaged trees have been removed or felled to the ground. Broken-off tops, severed large limbs, uprooting, and trunk wounds where more than a square foot of bark has been peeled away are serious injuries. Trunk wounds are common and unavoidable along the sides of skid roads. These "rub" or "bumper" trees should be left standing during logging operations so that the same trees will be skinned over and over again; usually they are removed at the end of an operation.

A Warning

In deciding to harvest trees, a landowner agrees to significant changes in the appearance of a forest. Frequently landowners are unprepared for the visual outcome of logging. Although a logging operation may have improved timber or maple sap productivity and greatly increased wildlife, an owner may have serious regrets.

Most often the focus of an owner's reaction is the appearance of slash on the site. However, this is usually a relatively short-lived effect, especially where the slash has been lopped close to the ground in partial cuts; the environmental impact is largely confined to unsightliness. In hardwood stands some slash can be eliminated in a separate operation to remove tops for fuelwood. Slash plays a role in the recovery of a forest after logging, since most of the nutrients in a tree are stored in the branches and twigs and are therefore recycled to the soil as the slash rots. Slash can also serve to soften the impact, and therefore the erosive effect, of water falling or running on disturbed soil.

Damage to residual trees, soil erosion, and water pollution are the most serious potential ecological and visual effects of logging. Stumps cut low, care taken to minimize damage to residuals, and all measures to reduce soil erosion and water pollution will minimize the visual impact of a harvesting operation.

A neat logging job is not necessarily the equivalent of a good job. Slash may be cut to the ground, while the quality of the residual stand has been lowered by highgrading. An excessive percentage of the land area may have been disturbed through poor road layout. The visual, environmental, and silvicultural aspects of a job must each be evaluated independently in order to arrive at a final judgment on the quality of the harvest.

Selling Sap

Maple sap is another saleable forest product. The sale of sap gathered by the landowner usually requires only a brief buyer-seller relationship, and only a brief written agreement on price. However, any arrangement in which the tapping, gathering, or sugarbush improvement work is to be done by another is best governed by a contract between the landowner and the person who rents the sugarbush.

Persons interested in buying sap or renting a sugarbush are much fewer than timber buyers; finding a buyer or renter might be difficult. In each New England state, except Rhode Island, there is a maple producers' association that might be helpful in finding a buyer for sap or someone to work a sugarbush under a leasing agreement. These associations are listed in appendix 4. As with hiring a logger, the credentials and experience of a person interested in buying sap or leasing a sugarbush should be reviewed, especially if he or she wants to assume the responsibility for silvicultural work to improve the sugarbush. In considering a prospective buyer or renter, the type of sap-gathering equipment to be used is not an important factor. Tubing systems and buckets have an equal impact on the trees. Vacuum pumps used to increase the sap flow in tubing systems can increase sap production by 70 percent or more, but are likely to have an adverse effect on tree vigor.

In the sample contract for the lease of a sugarbush that follows, and in the annotations to the contract, some of the special characteristics of harvesting sap are brought out.

Section I

Contract Date, Naming of Parties, Sugarbush Location,
Contract Expiration and Renewal.
1. This lease is made this __(date)__ of __(month)__ , __(year)__ , by and between __(landowner's name)__ of __(town, state)__ , hereafter called the landowner, and __(lessee's name)__ of __(town, state)__ , hereafter called the renter.

The renter or sap buyer is usually a commercial syrup producer or a middleperson who sells to a commercial producer.

2. The landowner, in consideration of the agreements with the renter set forth in this lease, hereby leases to the renter, to use for tapping to gather maple sap and to transport sap to a processing facility, the specific area located in the town of __(town)__ , in the state of __(state)__ , and described as follows:_____

_____

_____ .

The description of the area should include the road on which the sugarbush is located and the directions for finding the sugarbush from the road. This information should be followed by boundary data and a general description of the sugarbush; for example: "Twenty acres bounded on the east by town road No. 15, on the north by a stone wall, on the south by a red-blazed line, and on the west by a barbed wire fence, and containing approximately 1,200 sugar maple trees 40 to 120 years old, 10 inches to 30 inches in diameter at breast height."

3. This lease shall be effective on the __(date)__ of __(month)__ , __(year)__ , with the renter having the option for renewal for __(number)__ year(s) after the first year. The renter shall advise the landowner of intent to renew the lease by __(month)__ , __(year)__ .

Sap usually begins to flow around early March in southern New England and mid-March in the north. A contract must allow ample time in advance of the flow for tapping and setting up equipment; February 1 is a common date to begin a contract agreement.

With the advent of tubing collection systems, renters often require contracts that last five years or more. Tubing is cut to accommodate the particular arrangement of trees in a sugarbush, and is therefore not easily refitted to another maple stand. The expected useful life of tubing is about five years. For buyers who use buckets the duration of the contract may be less important, and a one-year renewal option may be satisfactory.

Section II

Approved Tapping and Management Practices
1. It is the intent of both parties that the maple trees on the land covered by this contract shall be maintained in their present condition, or improved, and the yield of sap maintained or increased with methods of tapping or tree thinning recommended in the landowner's sugarbush management plan, and that all roads and trails be left in good condition.

When a long-term agreement is made with a renter, he or she may be willing to perform cultural work, such as thinning in the sugarbush, or may even require this responsibility as a condition of the agreement. In such a case the landowner should have a sugarbush management plan for the area which the buyer must follow, or an understanding of good sugarbush management practices. If the buyer is just tapping and is not involved in cultural work, clauses such as this one are unnecessary, except for the stipulation regarding the restoration of roads and trails. Sugaring season is also likely to be mud time in New England; roads may be unavoidably damaged during sugaring operations and should be repaired as soon as sugaring ends.

If the landowner does not have a management plan which covers management of the sugarbush, other guidelines can be spelled out in the contract. For example: "thinning and other cultural work must be in accordance with the recommendations of the Extension Service and the state department of forestry," or, "in accordance with the guidelines in (*name of publication*)."

2. The following practices are mutually agreed upon:

A. No trees less than 10 (ten) inches diameter at breast height (DBH) shall be tapped. Only 1 (one) taphole 2 (two) inches deep shall be made in trees 10 (ten) to 12 (twelve) inches DBH; 1 (one) taphole 3 (three) inches deep in trees 13 (thirteen) to 18 (eighteen) inches DBH; 2 (two) tapholes 3 (three) inches deep in trees 19 (nineteen) to 24 (twenty-four) inches DBH; 3 (three) tapholes 3 (three) inches deep in trees 25 (twenty-five) to 30 (thirty) inches DBH; and 4 (four) tapholes 3 (three) inches deep in trees 31 (thirty-one) inches DBH or larger. No more than 4 (four) tapholes shall be made in any tree.

The vigor of trees smaller than 10 inches DBH is likely to be significantly diminished by tapping, since tapholes are about three inches deep and will penetrate a good percentage of the diameter of a small tree. The only trees smaller than 10 inches that may be tapped are those which have been selected for thinning in a young stand. Since they soon will be removed, it makes sense to harvest whatever sap they will yield. Although some older publications recommended more than four tapholes per tree per year, the practice is now thought to be excessively damaging.

B. Trees shall not be tapped over after a long period between sap flows.

A tree's defense system reacts to any injury, including a taphole, by gradually sealing it off. As it does so, the sap yield diminishes and eventually stops. This clause prevents the retapping of trees in the same season to restart the sap flow. Such a practice amounts to overtapping.

C. Chemical sanitizing agents in the form of paraformaldehyde pellets may not be used.

Applied directly to a taphole, sanitizing agents kill bacteria that would eventually fill it with a gummy substance and halt the sap flow. Although they can provide better quality sap and can keep a taphole open longer, paraformaldehyde pellets kill living tissue in a tree and encourage wood-decaying fungi. Other means of sterilization, such as submerging spouts in a clorox solution, are more advisable.

D. Thinning of maple trees to improve sap production by removal of the poorest-producing trees may not be done without first testing the trees for sap sweetness to determine their relative sweetness. Thinning will be done in accordance with the landowner's management plan. The number of trees to be removed will be determined by a forester, who will mark the trees to be cut with paint at breast height and stump height.

The procedure for testing sap sweetness is explained in detail in references listed in appendix 2. Any wood of fuelwood quality or better cut during cultural work is usually considered the property of the renter.

E. Sap-gathering vehicles such as tractors, sleds, or trailers shall be driven with care to prevent damage to tree trunks and roots.

Section III

Payment
1. The renter shall pay to the landowner the sum of _____ per taphole. Half of the payment is to be made (day and month) , (year) . The other half of the payment is to be made by (day and month) , (year) .

The rental fee per taphole is often established as 1 percent of the local retail price of a gallon of syrup, then adjusted upward or downward for unusually high or low sweetness of sap, or downward for difficult access. Average sap sweetness is about 2.5 percent sugar.

Rental payment is often remitted just after the end of the season, usually by May 1, with the balance due six months later. Any payment schedule agreeable to both parties is possible.

Section IV

Renter's Liabilities
1. The renter shall assume all responsibility and liability for accidents occurring to him/her, to his/her employees, or to visitors, while engaged in the tapping of trees, gathering of sap, thinning of maple trees, or cutting of wood on the area

covered by this lease, or while crossing any other lands belonging to the land-owner in the process of going to or coming from the area covered by this lease.

2. The renter shall be responsible for suppressing forest fires which may start while he/she is working on this property, and shall repair damage which occurs to roads or fences.

3. The renter shall watch for any evidence of insect, disease, or rodent damage which may occur in the area, shall advise the landowner if he/she feels action should be taken to control such damage, and shall advise the state department of forestry of especially threatening insect and disease infestations.

Section V

Landowner's Responsibilities
The landowner agrees to:
1. Furnish the area described above.

2. Pay all taxes and assessments on the real estate.

3. Allow the renter access to the areas described, as well as the use of existing roads for the purposes of tapping trees, hauling sap, or cutting and removing wood.

4. Keep cattle, sheep, and other livestock out of the area described above.

Livestock should be excluded from sugarbushes (and any productive woodland) because they consume and damage regeneration, wound tree trunks, and compact the soil. It is a mistaken belief that livestock manure on the forest floor contributes significantly to forest soil fertility.

5. Include the provisions of this lease in any deed of sale of this land to another party so that it will be binding upon the new owner.

This clause ensures that the contract remains in effect if the sugarbush is sold.

Section VI

Renter's Responsibilities
The renter agrees to:
1. Follow the landowner's management plan and approved management proce-dures for the development of existing young maple trees into tappable trees, and to protect those trees from damage.

2. Furnish all labor, equipment, and supplies and all operational expenses un-less use of the landowner's equipment is specified elsewhere in this agreement.

3. To pay any tax which may be due or become due on the sap or wood that is harvested.

In some New England states, a yield tax is levied on forest products. The concept of a yield tax is explained in chapter 7; the state laws requiring such payments are described in appendix 3. This clause does not exempt the landowner from liability for the income tax due on sap or wood payments. Income from the sale of sap is taxed as regular income.

4. Neither assign this lease to any person or persons nor sublet any part of the real estate for any purpose without the written consent of the landowner.

This clause prohibits transferral of the lease to another buyer without the owner's permission.

Section VII

Landowner's Right of Entry
The landowner or anyone designated by him/her shall have the right of entry at any time to inspect his/her property and/or the tapping, sap gathering, wood cutting, or other methods used.

Section VIII

Contract Contingencies
1. Failure of either the landowner or the renter to comply with the agreements set forth in this lease shall make him/her liable for damage to the other party. Any claim by either party for such damages shall be presented in writing to the other party, at least ten days before the termination of this lease.

2. If either or both of the parties to this lease die during the term of this lease, the provisions of this lease shall be binding on their heirs, executors, administrators, and assigns of the party or parties involved.

3. Any disagreements between the landowner and the renter shall be referred to a board of three disinterested persons, one of whom shall be appointed by the landowner, one by the renter, and a third by the two thus appointed. The decision of these shall be considered binding by the parties to this lease. Any costs for such arbitration shall be shared equally between the two parties to this lease.

Section IX

Cost-Share Payments
1. Any federal cost-share agreements are to be enacted jointly by the landowner and the renter. The renter is to receive _____ percent of the cost-share payments received from the U.S. Government for improvement done on the lease area.

Federal cost-share programs are administered by the U.S. Department of Agriculture; they are described in chapter 3. If the renter undertakes the woodswork in a sugarbush, he or she usually receives most or all of the government payments for approval practices.

Section X

Dated Signatures of Landowner, Renter, Witnesses

The Importance of Contracts

A contract should *always* be used in selling trees or leasing a sugarbush. It serves as a guide to standards for the work to be performed, as well as the liability of each party. In the event of a serious dispute or lawsuit, interpretation of the contract is the basis for determination of the outcome.

Any proposed contract should be reviewed by a landowner to make sure that all provisions are comprehensible and consistent with expectations. Anything that is unclear should be discussed with the supervising forester, or directly with the logger or renter if no forester is involved.

While a contract is an absolute necessity in any sale, it is no substitute for effective communication among landowner, forester, and logger. An owner who expects a high level of performance from foresters and loggers must assume the role of primary decisionmaker and provide clear direction early in the process of selling forest products.

7 Financial Aspects of Forest Management

This chapter describes the costs, returns, benefits, and risks of forestry investments. It also outlines the two methods most often used to evaluate and compare investment decisions concerning private woodland; identifies tax liabilities in forest management; and describes records that must be kept and those that it is advisable to keep.

Costs

Forest land, as an investment, has two components: the land itself, and the trees that grow on it. It is true that cutting trees may affect land values, and that the two components cannot always be dealt with separately. However, investments in land are not specifically treated here; the focus is on investments in the management of the trees for saleable forest products, namely, wood and maple sap or syrup.

The costs directly associated with forest management usually include foresters' fees, labor for timber stand improvement, the purchase and maintenance of equipment, and surveying associated with a timber sale or other management activity. Less common and less frequent costs include labor for planting or seeding, the purchase of seed or seedlings, the construction and maintenance of buildings, the purchase of easements, interest on loans, finance charges, premiums for liability and loss insurance, travel and office expenses, and measures to control insects, diseases, and fire.

Characteristics of Forestry Investments

Forestry investments have some distinctive characteristics not typical of other types of investments. Perhaps the most important is their time horizon. Investments in timber and sugarbush management may require up to fifty years to generate a net return. During that period only a fraction of the costs, at best, may be recoverable; unlike some other investments, there is usually no option for early liquidation at a reduced rate of return until the trees are ready for harvest or tapping. The time factor can effectively reduce the liquidity of a forestry investment.

After a stand has grown to merchantable or tappable size, liquidity increases markedly because the value of the wood or sap is efficiently stored "on the stump" for a relatively long time before the stand begins to deteriorate and lose value. (During this time, in the case of standing timber, value is compounding *without an annual income tax liability*—a very advantageous characteristic of timber investments.) Liquidation can therefore often be timed to meet an owner's cash needs or to take advantage of upward price trends, as well as to benefit the stand silviculturally. Finally, installments sales of timber can help to spread income over several years, which has income taxation advantages for some owners.

Components of Investment Return

The return from a timber management investment is determined by changes in tree values; these occur in three ways. First, wood markets fluctuate—that is, stumpage prices vary regionally and nationally over time. Historically, the growth in prices paid for sawtimber has exceeded the inflation rate, but this may or may not hold true for other products, over short periods of time, or for individual species. When the rate of return does not surpass the inflation rate, the investment in timber management will break even or may fall below the rate of return for similar long-term investments. Because of the long horizon of forestry investments, the accuracy of an estimate of future timber prices is critical, and the projected prices should be periodically reviewed. Technological developments will continue to make it possible to substitute less expensive (more plentiful) timber species and sizes for those now used for certain products, so projections are especially tricky. Increases in what we will call the *market value*, then, depend on factors not immediately related to forest management, which are therefore beyond the scope of this discussion.

The second way in which trees increase in value can be greatly strengthened by timber management: this is *volume increase*. A tree or a stand grows in value as it grows in size because it adds more board feet, or cords, or cubic feet, or tons. In financial analysis, volume increase is usually expressed as a rate of growth. As a forest stand grows each year, wood fiber—volume—is added. Each year's growth is the "interest" added to the "principal," which is represented by the volume that has already accumulated. The growth rate is the percentage of the existing volume represented by the new growth. It is usually estimated on the basis of the number of years a tree is currently taking to add 1 inch of radial growth; that is, on the number of rings per outside inch of an increment core. Table 22 charts the growth rates of trees of various diameters with various numbers of annual rings per inch.

Despite the best management, as trees get larger the growth rate usually declines because the volume already accumulated is so large that the proportion represented by new growth shrinks. Also, as trees achieve maturity their growth begins to slow. Table 22 shows that a 6-inch diameter tree growing 5 rings per inch has more than a 14 percent growth rate, while a 12-inch tree growing the same number of rings per inch has only a 7 percent growth rate. If volume increase was the only component of value growth, a purely financial outlook would dictate cutting a tree when its growth rate to date or below the so-called "alternate rate of return"—the interest rate that could be earned by other available investments.

However, the eventual decrease in the rate of return resulting from volume changes is usually more than offset by an increase in *unit value*, which is the third way that trees change in value. The unit value is the worth of each cord, cubic foot, or board foot that is added as the tree grows; it increases as the tree increases in size. As we have seen, log size (especially diameter) is preeminently important in grading logs; products of higher value can be obtained from larger trees than from smaller ones, assuming the quality of trees to be equal. Larger trees are cheaper to harvest and saw, and they yield wider boards with more clear lumber (fig. 77). A cubic foot of wood from a 20-inch tree may be worth several times as much as a cubic foot from a 10-inch tree of comparable quality. And although there may be little change in height or diameter, the growth of a 20-inch tree means a great increase in volume and quality.

Changes in volume and unit value depend on species, site productivity, individual tree quality, and stand density. So unlike other investments in which interest yields a steady increase in value, the appreciation of timber investments are more steplike. Steep value increases occur when trees change in product potential. For instance, when trees change from cordwood to sawlog size there can be as great as a 1000 percent increase. When high quality trees become large enough to be used for veneer another 200 percent increase is possible. Management can maximize the

Table 22 Rates of Tree Growth

Tree Size (DBH in inches)	Rings per Inch											
	4	5	6	7	8	9	10	11	12	13	14	15
	Growth Percentage											
4	28.2	22.5	18.8	16.1	14.1	12.5	11.3	10.2	9.4	8.7	8.0	7.5
5	22.0	17.7	14.7	12.6	11.0	9.8	8.8	8.0	7.3	6.8	6.3	5.9
6	18.1	14.5	12.1	10.3	9.1	8.0	7.2	6.5	6.0	5.6	5.2	4.8
7	15.3	12.3	10.3	8.8	7.7	6.8	6.1	5.6	5.1	4.7	4.4	4.1
8	13.3	10.6	8.9	7.6	6.7	5.9	5.3	4.8	4.4	4.1	3.8	3.5
9	11.6	9.3	7.8	6.7	5.8	5.2	4.7	4.2	3.9	3.6	3.3	3.1
10	10.5	8.4	7.0	6.0	5.3	4.7	4.2	3.8	3.5	3.2	3.0	2.8
11	9.5	7.6	6.3	5.4	4.7	4.2	3.8	3.4	3.2	2.9	2.7	2.5
12	8.7	7.0	5.8	5.0	4.4	3.9	3.5	3.2	2.9	2.7	2.5	2.3
13	8.0	6.4	5.3	4.6	4.0	3.5	3.2	2.9	2.7	2.5	2.3	2.1
14	7.4	5.9	4.9	4.2	3.7	3.3	3.0	2.7	2.5	2.4	2.1	2.0
15	6.9	5.5	4.6	3.9	3.4	3.1	2.8	2.5	2.3	2.1	2.0	1.8
16	6.4	5.2	4.3	3.7	3.2	2.9	2.6	2.3	2.1	2.0	1.8	1.7
17	6.1	4.8	4.0	3.5	3.0	2.7	2.4	2.2	2.0	1.9	1.7	1.6
18	5.7	4.6	3.8	3.3	2.9	2.5	2.3	2.1	1.9	1.8	1.6	1.5
19	5.4	4.3	3.6	3.1	2.7	2.4	2.2	2.0	1.8	1.7	1.5	1.4
20	5.1	4.1	3.4	2.9	2.5	2.3	2.0	1.9	1.7	1.6	1.5	1.4
21	4.9	3.9	3.3	2.8	2.4	2.2	2.0	1.8	1.6	1.5	1.4	1.3
22	4.6	3.7	3.1	2.7	2.3	2.1	1.9	1.7	1.5	1.4	1.3	1.2

Source: B. S. Ashley, *Reference Handbook for Foresters* (Morgantown, W. Va.: U.S.D.A. Forest Service, State and Private Forestry, 1980), p. 17.

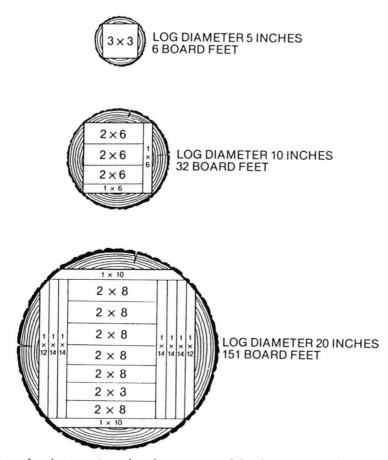

Fig. 77. Lumber quantity and quality improve with log diameter. An 8-foot log with a small-end diameter of 5 inches (top) yields ⅕ as much lumber as a log of the same length with a 10-inch small-end diameter (*center*), and only ¹⁄₂₅ as much as a 20-inch log of the same length (*bottom*). In addition, the large log will yield wider (more valuable) lumber that will usually have a larger proportion of knot-free boards.

positive contribution of each of these factors to tree value by applying appropriate techniques:

1. Intermediate treatments and regeneration harvests work to control the species composition of a stand, and to encourage the most valuable and discourage the least valuable species.

2. Thinnings and improvement cuts allocate volume increases to the highest-quality trees in a stand, improve their growth rate, and improve productivity by capturing the volume of overtopped trees that would eventually die. Although a stand that is overstocked will produce the same amount of wood fiber as a stand with optimum density, the increase in diameter growth of each tree will be slower, and volume will be lost in trees that succumb to suppression.

3. Planting (where it can be used economically) improves species composition in a stand, and supplements understocked stands in which volume increase would otherwise not be maximized (in such stands, each tree is growing at a fast rate, but the land is capable of growing more wood by carrying more trees). Planting is still an expensive and rare practice.

For maple sap production, the variables of market value, volume, and unit value are also important as the elements of investment return. Again, the market value of maple products is not directly a function of forest management. The price of maple products has followed the price of food, with a few exceptions, for about the last sixty years. The volume of sap available in a sugarbush can be maximized through management by maintaining full stocking of fast-growing sugar maples of tappable size. The highest unit value can be realized by favoring those trees with the highest sap sugar content.

Risks

There are two kinds of risk associated with a forest management investment. One is what might be called "physical" risk: the prospect of value lost to insects, diseases, fire, or other natural catastrophies. The extent of this risk is highly variable, and depends on location within New England. For instance, stands of spruce and fir in Maine are highly susceptible to extensive damage during spruce budworm outbreaks, which occur from about every forty to eighty years; such outbreaks make investment in these species risky. In recent decades, forest fire has been a low risk in much of New England. Although commercial insurance against natural catastrophe is unavailable to most private landowners, there are many management practices that can reduce physical risk. A good road system, for example, permits effective fire control, and careful treatment of slash reduces the risk of fire. Maintaining stands of trees in a vigorous condition reduces the threat of attack by many insects and diseases.

The second type of risk is associated with the fluctuations of the markets for wood products, which subject is quite complex and beyond the scope of our discussion.

Evaluating Investments in Management

A very common approach to evaluating the investment status of a forest stand is to compare its current growth in value with a chosen alternate rate of return (adjusted to exclude inflation), such as that available from money markets or savings accounts. When taking this approach, which is termed the *financial maturity* method, it is time to liquidate the management investment (regenerate the stand) when the rate of value increase drops below the chosen alternate of return. This method of evaluating does not consider past land purchase or forest management costs, but looks only at the current rate of stumpage value increase.

A second common approach to evaluating investments, the *discounted cash flow* method, seeks to establish the present value of anticipated future returns. Because costs are incurred and revenues are realized at different times in the future, the method is used to adjust these costs and revenues so they can be compared at a particular time, usually the present. Such an analysis requires the selection of a minimum acceptable interest rate, the choice of an investment period, and estimates of future costs and revenues. The process is best illustrated by an example. We will take the case of a 20-year-old northern hardwood stand, 4 inches DBH. The problem is to determine the financial advisability of investing $80 per acre now in a girdling operation for timber stand improvement (TSI). Will this investment increase the total value of the stand over a chosen investment period (30 years) at the 4 percent rate chosen as the alternative real rate of return?

The first step is to compare projected costs and revenues without and with the investment in the girdling operation; table 23 includes a summary of the costs and returns in this example. The TSI treatment accomplished by girdling results in an increase in the diameter growth rate of the stand. As a result, a fuelwood thinning, yielding an anticipated $36 per acre, would be possible in 10 years at age 30. By age 50, two additional thinnings would be made, resulting in a total yield of $200 per acre. Without the investment, stand growth would be much slower, and the first thinning would not be possible until age 40; by age 50, the total expected yield from thinnings would be $100 per acre.

The second step is to determine the present value of each anticipated revenue. The first thinning yields a revenue of $36 per acre; the problem is to determine what value $36 represents today, using the 4 percent interest rate selected. If the investment is made in TSI, the present value of $36 is $24.32. Another way of phrasing this would be to say that $24.32 invested

Table 23 Costs and Revenues for TSI Investment Analysis

Year	Stand Age (years)	Growth Rate (rings per inch)	Avg. Stand DBH in inches	Operation/Yield	Cost (dollars)	Revenue (dollars)	Present Value[a]
				Without TSI			
0	20	20	4	None	—	—	—
10	30	20	5	None	—	—	—
20	40	20	6	Fuelwood thinning; yield 6 cords at $6/cord	—	36.00	16.43
30	50	10	8	Fuelwood thinning; yield 8 cords at $8/cord	—	64.00	19.74
				Total	—	100.00	36.17
				With TSI			
0	20	20	4	TSI (girdling operation)	(80.00)	—	(80.00)
10	30	10	6	Fuelwood thinning; yield 6 cords at $6/cord	—	36.00	24.32
20	40	10	8	Fuelwood thinning; yield 8 cords at $8/cord	—	64.00	29.21
30	50	10	10	Thinning; yield is 2,500 BF pallet logs at $40/MBF	—	100.00	30.84
				Total	(80.00)	200.00	4.37

[a]Discount rate is 4 percent.

for 10 years at 4 percent compound interest would yield $36. This calcula-
tion is completed for each anticipated revenue: the net present value of the
two options is $4.37 if TSI is undertaken, and $36.17 if it is foregone.
From a purely financial point of view, the TSI investment would not be
justified in this example. Federal cost-share programs may significantly
reduce the owner's out-of-pocket expense. If the landowner represented in
table 23 were eligible for federal assistance, then the cost of the TSI would
have been lowered to approximately $28. The net present value would
have increased to $63.37 making the TSI a good investment.

However, other factors not included in the example might lead an
owner to make the TSI investment without the federal cost share funds.
One factor might be the promise from the TSI of a shorter rotation and
higher-quality timber when the stand is finally harvested. Although length-
ening the investment period to include an entire rotation increases the effect of the interest rate, the improved value of the stand attributable to the
TSI might make the investment economically advisable. Finally, it is
worth repeating that woodland owners in New England rarely have purely
financial goals in undertaking forest management.

It is important to note that the discounted cash flow method increases
the attractiveness of investments that yield revenues in their early years
and incur the bulk of expenses in their later years. Opportunities for early
revenues from forest management will become more frequent as a trend
toward the utilization of increasingly smaller trees continues.

Computations for the two methods of financial analysis described involve use of a few formulas and compound interest tables. *The outcome is
always strongly affected by the choice of an interest rate, or alternative rate
of return, because of the long time periods involved.* Most foresters can
furnish the information needed to make financial analysis calculations.

In general, landowners considering an investment in forest management
practices should concentrate on stands of valuable species that have the
potential to respond to treatment and are on good and accessible sites.

Taxes

Taxes can significantly color the forest management investment picture.
Four types of taxes potentially affect the forest landowner: *income, property, yield,* and *estate.*

Table 24 Recovering Costs Related to Forestry

Cost	Tax Structure			
	Hobby	Investment	Passive Business	Active Business
Non-capital management expenses[a]	Deduct expenses in years where there is income from property	Deduct expenses as long as they exceed the minimum miscellaneous deduction (Schedule A)	Expenses carried forward to a year when there is property income	Deduct expenses in the year they occur
Capital expenses (non-depreciable)[b]	Recover expenses when land or timber is sold	Recover expenses when land or timber is sold	Recover expenses when land or timber is sold	Recover expenses when land or timber is sold
Capital and non-capital expenses (depreciable)[c]	Deduct depreciable amount only in years where there is income	Deduct depreciable amount as long as it exceeds the minimum miscellaneous deduction (Schedule A)	Depreciable expenses can be carried forward to a year when there is property income	Depreciation can be taken each year throughout "useful" life of the product
Capital expenses that are depletable	Recovery at time of timber sale	Recovery at time of timber sale	Recovery at time of timber sale	Recovery at time of timber sale
Property taxes	Deductible annually	Deductible annually	Must be carried forward to a year when there is property income	Deductible annually
Reforestation	Not applicable	Tax credit available if minimum miscellaneous deductions met	Must be carried forward to a year when there is property income	Credit available in the year reforestation takes place

[a]Non-capital management expenses include all costs related to a timber sale, equipment with a life of less than one year, and minor repair of equipment.
[b]Non-depreciable capital expenses include all costs associated with the acquisition of property, property rights, or permanent improvements like improving roads.
[c]Capital expenses that are depreciable include land improvements such as bridges, culverts, or fences. Non-capital expenses that are depreciable include necessary equipment with a life greater than one year and major repairs of equipment.

Income Taxes

Although the profit realized from the sale of forest products is taxable, tax liability can be minimized. Choosing the most appropriate and advantageous forest land tax structure is the first step. The structure choices for forest owners are hobby, investment, passive business and active business. They vary chiefly according to their eligibility standards, landowner involvement and "cost recovery" advantage (table 24).

A small percentage of forest owners in New England will find that their forest activities must be considered a hobby; that is, these activities are not conducted with the primary intent of realizing a profit. It is the easiest tax structure to apply but, unfortunately, it provides the least reduction in taxes. Management expenses cannot be carried forward and can be deducted only in a year that income is earned from the property.

The investment tax structure is suitable for forest land activities that are undertaken to realize a profit and in which the owner is minimally involved. Landowners with other investments may be in this category. Landowners who qualify for this structure can deduct expenses in the years they occur, as long as the costs from all investments exceed the minimum "miscellaneous deduction" amount which is a fixed percentage of adjusted gross income.

Passive business income is defined as that generated by a business in which the landowner does not "materially participate." That is, the forest owner does not meet a series of rigorous tests designed to prove significant work involvement. In this category, expenses can be carried forward to a year when the property is producing income and then deducted. Unlike the investment structure it is not necessary for costs of a passive business to exceed a fixed percentage of the owner's adjusted gross income.

The most advantageous tax structure is the active business because it provides unrestricted deductibility of expenses. Landowners who qualify for this structure must pass the tests which prove the owner is actively engaged in the business. All non-capital costs can be deducted or depreciated in the year they are incurred regardless of whether income from the property is generated. Most landowners can utilize this category if they are willing to invest the time necessary to set up and maintain a business.

The timber depletion allowance, capital gains treatment, depreciation, reforestation tax credit, capitalizing and expensing are the available means of minimizing tax liability in timber operations. The extent to which each is available to a landowner depends on the tax structure (table 24). Note that the net revenues from maple sap and maple products are ineligible for depletion and capital gains treatment and are taxed as ordinary income. Timber, however, is considered a capital asset and therefore receives a special tax treatment.

The timber depletion allowance.

In the liquidation of most investments, profit can be neatly calculated as the difference between the cost of the investment and the sale proceeds. The profit on a timber investment is difficult to calculate, however, because the investment is usually not liquidated at a single point in time. The timber depletion allowance is designed to accommodate this situation, and allows an owner to gradually recover a timber investment. It is not really a "depletion" in the sense ordinarily used in taxation; it is really just a means of getting back much of the money spent on the timber. It is applicable to the sale of timber whether the ownership is considered a hobby, investment or a business.

The use of the timber depletion allowance is best illustrated by an example. The first step in its calculation is to separate the investment in the land from the investment in the timber, because the timber depletion allowance is formulated to recover capital investment costs in the timber only. For our example, we will postulate a woodland tract that was purchased for $23,000. Additional costs of acquiring the land were those for a title search and filing, a survey, travel, and a forester's evaluation of the standing timber; additional costs totalled $2,000. The total cost of acquiring the land, then, was $25,000. The forester determined that there were 200,000 board feet of timber on the land when purchased and estimated its fair market value to be $6,000. The current fair market value of the land, which is the value estimate made for property tax purposes and is not necessarily equal to the sale price of a piece of property, was $21,500. Of the original $25,000 investment, then, $7,000 was attributable to the timber:

$$\$6,000 \div \$21,500 = 28\%$$
$$\text{and}$$
$$28\% \text{ of } \$25,000 = \$7,000$$

This method of allocating part of the purchase price to the timber is prescribed by the Internal Revenue Service, and is further explained in references listed in appendix 2.

In the example, the $7,000 cost of the timber is the basis to be recovered, or "depleted," as the timber is sold. The basis will be adjusted as further investments are made in timber management and as timber is sold.

Ten years after the land purchase some of the timber was ready for sale. At this point, the forester determined that the total timber growth on the property since its purchase had been 50,000 board feet, so there were now 250,000 board feet:

$$
\begin{array}{r}
200,000 \text{ bf volume at time of land purchase} \\
+50,000 \text{ bf growth, next ten years} \\
\hline
250,000 \text{ bf current volume}
\end{array}
$$

The owner's investment in the current timber volume was 2.8 cents per board foot, more conveniently expressed as $28 per thousand board feet:

$7,000 initial investment (basis) ÷ 250,000 current volume =
2.8¢ per board foot, or $28 per thousand board feet

The unit value of $28 per thousand board feet represents the owner's investment in a thousand board feet of the timber to be sold in the current sale; it is known as the "depletion unit." At each sale in the future the adjusted basis divided by the current timber volume will be recalculated to determine the depletion unit for that sale.

From this current volume, a sale of 100,000 board feet is made. Its investment value is $2,800:

$$
\begin{array}{r}
100,000 \text{ board feet} \\
\times \$28 \text{ depletion unit} \\
\hline
\$2,800
\end{array}
$$

This sum represents the portion of the original basis ($7,000) being liquidated.

Gross revenues from the sale are $8,000, so the owner's profit, using the timber depletion allowance, is $5,200:

$$
\begin{array}{r}
\$8,000 \text{ gross revenues} \\
-2,800 \text{ investment value of timber sold} \\
\hline
\$5,200
\end{array}
$$

In calculating the income tax from timber sales, the costs of selling the timber can usually be deducted from the proceeds as long as the costs are incurred in the same tax year as the income received. If the owner in the example paid the forester on an hourly basis a total of $1,500 to mark and administer the sale, that sum would have been used to further reduce the taxable profit to $3,700:

$5,200 gain using depletion allowance
– 1,500 costs of the sale
_____
$3,700

By using the timber depletion allowance and deducting the costs of a sale, the owner reduced the taxable sale income from $8,000 to $3,700. The owner recovered or "depleted" $2,800 of the original investment of $7,000 (the basis). In future sales $4,200 will be the adjusted basis that remains to be recovered.

Capital gains treatment.

Under certain circumstances, profits on timber sales are considered capital gains and are eligible for special tax treatment. The extent of this benefit has varied greatly over time. Until 1986, capital gains treatment meant that only 40 percent of the gain was taxable. After 1987 all of the gain was taxable but not subject to social security tax. It is very important for landowners to track the changing capital gains benefits through consultation with a forester or accountant. The attractiveness of this tax treatment is likely to increase again in the next decade, possibly resulting in significant income tax savings. In the example above, if the sale had taken place in 1985 the capital gains treatment would have further reduced taxable income from $3,700 to $1,480. Combining all of the available tax deductions including depletion allowance, expensing, and capital gains, the owner would have been able to reduce the taxable income from $8,000 to $1,480.

The conditions under which income from timber has been eligible for long-term capital gains treatment are easily met by most private woodland owners. In very simplified form, they are: (1) the timber sold had been held for at least one year; and (2) the owner had chosen to structure the income from the forest land as hobby or investment but not as passive or active business. In the eyes of the IRS, an owner who makes frequent lump-sum sales may be considered to be in the timber business, in which case that owner's sale proceeds are no longer eligible for capital gains treatment. The IRS has not defined what "frequent" means.

Depreciation.

Depreciation is the means to recover investments in equipment and in major improvements to the land that wear out over time. Each such ex-

penditure is assigned an "economic life" in each year of which a portion of the total cost representing wear and tear on the item is subtracted.
Only those who consider their forest land an active business can annually
receive depreciation on appropriate items without limitations. Owners in all
other tax structures are subject to restrictions which may even eliminate
the use of this recovery. For passive business, depreciable costs are recorded
each year but the loss gets "impounded" until the activity produces an income. For landowners in the hobby category, depreciable costs can only be
taken in a year that income is received. For the investment structure, if depreciable costs exceed the percentage floor of adjusted gross income, depreciation is allowable. Common depreciable investments for a small woodland
owner would include chain saws, pickup trucks, harvesting equipment,
bridges, culverts, graveling, temporary road construction costs, and sugaring
equipment. The income tax treatment of woodland roads is fairly complex,
and depends on purpose, premanency, and the extent of use. The IRS can
supply guidelines for the normal recognized life of depreciable investments.

Reforestation tax credits.

Although little tree planting or seeding for timber production is currently
undertaken in New England, a federal income tax credit for artificial regeneration is now available to woodland owners who have structured their
forest ownership to be an investment or business and may make the practice more economical. The law (PL95-451), passed in 1981, allows a 10
percent tax credit for artificial regeneration expenses. Direct costs can include site preparation, seeds and seedlings, labor (excluding that of the
owners), and depreciation of tools and equipment. The law allows the
owner to deduct these expenses from annual income over seven years, and
to subtract 10 percent of those costs from the taxes owed to the federal
government for the year in which the planting or seeding work was done.
Similar limitations for the use of this tax credit by the investor or passive
business apply here (table 24). The law limits eligible reforestation expenses to $10,000 per year, and the maximum size of eligible tracts to
1,000 acres per owner.

Treatment of timber expenses for income taxation.

In calculating the taxable profits from timber management, some expenses can be deducted from gross income in the year they were incurred.
This is called "expensing" a cost. The ability to expense a cost will again
depend on investment structure (table 24). Other costs—in general, those
that are incurred "for the acquisition, improvement or reforestation of an

asset having a useful life of more than one year"—must be "capitalized"; that is, they are added to the long-term investment value (basis) of the timber or land, or they are depreciated. If a cost is capitalized, it is only recovered when the timber or land is sold; it is added to the amount deducted from the sale proceeds to find the net profit.

The income tax aspects of forest management can be complex. This discussion has presented a simplified overview. The important point to be made is that landowners should use the special tax provisions for lowering their income taxes on forest products. Because the necessary tax forms, regulations, and record keeping seem puzzling or tedious, many owners disregard these provisions, and pay much more tax than is required. Although evading taxes is illegal, avoiding unnecessary taxation is wise.

Property Taxes

Property taxes are considered a land ownership cost rather than a management cost. For income tax purposes their treatment will again depend on the investment structure of income produced from the forest land. For hobby and investment income, taxes can be deducted as an itemized expense. The passive business can have this expense forwarded to a year when there is property income. While the active business owner can deduct property taxes as a regular business expense, most New England states allow the reduction of the annual property tax burden for woodland through participation in a so-called current use program (see appendix 3).

Yield Taxes

In some New England states yield, or severance, taxes are assessed against a landowner when forest products are harvested. They are generally calculated as a percentage of the gross value or sale price of the stumpage. See appendix 3 for an inventory of yield tax laws in New England.

Estate Taxes

Since 1987, the tax-exempt value of an estate is $600,000. Estates worth less can be passed to a spouse or other heirs tax free. For most landowners this means that land and timber can be transferred intact. Without proper planning, a higher estate value can result in the sale of the timber or land to meet estate taxes. Proper planning might involve, for example, passing $10,000 annually to each child, tax free.

For estates that exceed the $600,000 value the sale or donation of the land's development rights through a conservation easement to an appropriate non-profit group such as a land trust will lower the estate value. Conservation restrictions spell out the prohibited and permitted uses of the property. As a general rule they allow a full range of forestry, agricultural and other open space uses but limit or prohibit development. Each conservation restriction is tailored to meet the needs of the landowner. It may yield an income tax deduction based upon difference in the appraised value before and after the restriction. With regular monitoring by a land trust, such an arrangement will insure that the land is perpetually conserved and managed according to the current owner's wishes.

There are several other land conservation options with tax advantages for some landowners including gifts of lands to certain conservation groups, or "bargain sales" below fair market value to a land trust. These will probably yield income tax savings while insuring the management of the land subject to the conservation restrictions. See appendix 4 for a list of information sources on long term conservation agreements.

Because timber is relatively illiquid, land value alone is usually considered by banks in reviewing most mortgage applications and by appraisers who value land holdings for estate purposes. For owners with a high estate value that will result in tax liability for a surviving spouse or heirs, the fact that timber is often undervalued is advantageous.

On the other hand, for owners whose estate value is below the amount allowed to pass tax free to survivors, it is advantageous to avoid undervaluation of the estate and to have valuation be as close as possible to the tax-free allowance without exceeding it. The higher the estate value passed to a spouse or heirs, the higher will be the inherited basis. A high basis will make for a smaller taxable profit should survivors later sell all or some of the property included in the estate.

Like income taxation, estate taxation is complex, with many gray areas. This introduction should serve only as a stimulus to thinking and to talking to a tax adviser who is familiar with the peculiarities of forest ownership.

Cost-Sharing Programs

Many people are reluctant to make investments in forest management because of the long time periods involved. To encourage investments in management, the U.S. Department of Agriculture operates several cost-sharing programs. The agency makes a refund, sometimes tax exempt, of a percentage (usually from 50 to 75 percent) of the total cost of an approved activity. The management practices most often cost-shared are precommercial thinning, management planning, the pruning of softwoods, and tree planting. Eligible practices will vary from state to state and from year to year. The programs are funded through the Agricultural Stabilization and Conservation Service (ASCS). County foresters are usually responsible for on-the-ground implementation. An interested landowner should contact either the county forester or the ASCS office in the county in which the land is located.

Record Keeping

One conclusion to be drawn from the preceding discussion is that owners should keep records of all activities, costs, and revenues associated with forest ownership and management in all tax structures. From a financial viewpoint, records of four types should be kept: (1) a simple financial journal in which all expenditures and income are recorded; (2) ledgers of timber and land accounts, and perhaps building and equipment accounts, showing adjustments to each basis over time; (3) tax records; (4) a document file containing all bills, receipts, cancelled checks, invoices, and sale contracts. The information in such records is important for investment analyses, assessments of loss, demonstrations to the IRS of an owner's intentions as an investor, and tax computations.

Financial journal.

It is advisable to keep a single journal showing the date, amount, and description of all credits and debits related to management. This will serve as an index to and as supplementary information for the timber, land, and other account ledgers. Such a journal is also useful from both financial and silvicultural points of view for keeping records of all woodswork done on a property; this can be accomplished by using a short form such as that shown in figure 37. Information of this kind would be useful to a landowner who makes unit-priced timber sales and is asked to present proof to the IRS that an effort is being made to profit from timber management.

Ledgers of timber and land accounts.

Investment in a piece of woodland must be allocated to both timber and land accounts for taxation purposes, as indicated in the example used to explain the timber depletion allowance. Records should be kept of the basis in the timber account as it is adjusted over time—upward for capitalized costs and newly acquired timber, downward for timber sold or casualty losses. The basis in the land will be adjusted upward when pertinent capital investments are made, and downward when land is sold.

The timber depletion example also showed that it is necessary to keep a running account of total timber volume, as well as of timber value. It is highly advisable to set up these accounts as the land is purchased, and it is especially important to obtain an estimate of the timber volume at that time. If the original volume purchased with the property has to be estimated a long time after the purchase, it will be necessary to hire a forester to calculate the total growth over the period, and to deduct that growth from the current volume to arrive at the original volume. A field cruise done at the outset of ownership is a less expensive proposition. During the initial cruise at the time of purchase stand growth rates may be estimated, making subsequent updates of the timber volume account relatively easy.

Although the maintenance of timber and land accounts is most crucial at the time of timber sales, there are reasons to update them more regularly—ideally, every year. One important reason is that information is immediately available and current should it be necessary to file a casualty loss with an insurance company or the IRS.

Keeping track of land and timber accounts and timber volume is best done in a ledger in which all expenditures, income, and volume gained and lost can be recorded. In recording expenditures, it is important to show whether they were capitalized or expensed. Those that are capitalized will be recorded in the ledger; those that are expensed will appear in income tax records.

Tax records.

A record should be kept of all tax payments—income, property, yield, and estate. Income tax records should include details of any expensed items and any tax credits claimed, and copies of all tax forms filed. There is no kind, format, or content of records prescribed by the IRS other than Form T and its accompanying schedules. This form must be completed when timber is bought or sold, when a casualty loss is claimed, when reforestation or timber stand improvement has been accomplished, or when a timber depletion allowance is taken. Form T includes schedules for calculating the apportionment of costs between land and timber, for adjusting the timber basis, and for reporting timber sales and losses. In addition to keeping copies of Form T, it is advisable to retain detailed supplementary taxation records, especially those described in the other record categories outlined above.

Document file.

A landowner should retain all bills, receipts, and cancelled checks that show the income, deductions, and credits reported in tax returns. All timber sale contracts should be kept as well.

Postscript: On Stewardship

Misused for centuries and once virtually eliminated
from much of the region, New England's forest is resil-
ient enough to have reclaimed 80 percent of the region's land area . . . Other regions of the world have
not been so fortunate . . . Where trees have been cut
in large quantities in these regions, the fragile soils are
extremely vulnerable to erosion, and intense heat robs
them of nutrients . . . Under the desert conditions
that result, trees do not regenerate, and the social and
economic impacts are devastating and sometimes irreversible. In New England, our forests persist, and our
options are, for the time being, still open.

Ten years after concluding the first edition of this book with those op-
timistic words we are still amazed by the resilience of the New England
forest. We find ourselves less sure, however, about its future. Our confidence
has been eroded by a decade of observation and speculation regarding the
effects on the forest around us of air pollution, a series of new forest insects
and diseases, and unmanaged subdivision and development of woodland.
We worry, too, about the forest effects of global threats: climate change,
deforestation, ozone depletion, and the stunning increases in human population, all of which once seemed like issues far outside the stone wall
boundaries of rural New England. These issues now give us a common
bond with the tropical forests.

We now know that New England's quiet woods are not apart from the
environmental, social, and economic stresses bearing down on other areas
of the region, nation, and world. Thick air pollution blows into the Green
Mountains from Ohio. Harmful insects are transported to Connecticut from

Asia. Clearcutting in the Amazon and Malaysia, possibly warming the global climate, could threaten maple syrup production in the Berkshires. Corporate finance in London determines the fate of the large industrial forest ownerships that hold together the ecosystem of far northern New England.

Analogous to New York City's Central Park on a regional scale, New England's green forests are being subdivided by the need to escape gray cities of crime and pollution, or to find a house lot on affordable land. To many of those who crave a piece of the woods, ten acres seems a vast and wild refuge, but, ecologically and economically, such subdivision makes fragments too small and too numerous for practical wildlife, timber, or landscape conservation.

Skeptics might regard our concern as unfounded. The piece of forest history we provide in Chapter One of this book indeed shows that the New England forest has been shaped by external forces since the time of European exploration and settlement: seventeenth-century English kings high-graded the forests for naval masts and timbers. Much of New England was cleared to feed the cities of the new American republic in the 1800s. Foreign demand for oak claimed much wildlife mast in the 1970s. But the new forces from outside are not so easily understood or reversed: no war, emigration, or market shift will relieve these pressures, only concerted, intentional, and, in some cases, global efforts.

Others may fault us for lack of a scientific basis for our worry, but there is conclusive evidence for neither their confidence nor our concern. We believe that it may be the complexities of research in a forest ecosystem that make elusive any scientific certainty about the forest effects of, for instance, air pollution or ozone depletion, and not a lack of effects. Our concern is based in that belief and in common sense that tells us that the impact on forests of altered air, rain, climate, and sunlight is more likely than not to be negative and serious.

Based on our concern, we would redefine land stewardship. We once thought that stewardship was defined by the legal boundaries of a piece of property and by the conscientious application to it of scientific forest management techniques. But the lesson of the last decade has been that real stewardship is only partly a matter of technique. To follow the textbook practices and observe the forestry laws explained in this and other manuals is essential, but no longer enough.

We therefore now define stewardship as very cautious forestry that reflects an understanding that even a small piece of New England woodland is part of a bigger ecosystem that is regenerative, but precious and threatened. It is management that looks past legal property boundaries to understand the ecological whole of which any small woodlot is a part: Is this small corner of hemlock part of a contiguous deer yard? Is this beech stand a short stretch of a long route that bears use to travel to breeding grounds? Is this tiny trickle a headwater? Are these big trees a patch of rare "old growth"? Stewardship is a share of responsibility bigger than the acreage actually owned; it is an understanding that each acre is not "just woods" but a link in the forest's safe passage to generations of people, trees, and animals.

We used to think, too, that it was enough for an owner to *know* about good forestry, to understand what the forester and logger talk about, to make sure that the logging job was a good one. But stewardship is also about *looking*. No book or hired forester is a substitute for the instinct an owner can develop by simply watching the woods, year in, year out.

In a time when the beacon of scientific certainty is dim and environmental and social changes are unprecedented in their substance and scale, stewardship prefers smaller harvesting equipment to large, longer rotations to short, and uses clearcutting only with great care. It times harvests to avoid disruptions of nesting birds. It goes beyond legal requirements for protecting rare and endangered species and wetlands, which requirements, for reasons of enforceability and political expediency, are often minimal, or omitted altogether from forestry regulations. Stewardship is an ethic that well exceeds what is merely technically correct, acceptable, or enforceable.

Perhaps most important is that stewardship must have, in addition to a broader view of the forest, a longer sense of time than mere ownership. Consideration of the long-term future of a piece of woodland well before the current ownership ends, either by death or sale, is essential. A lifetime of careful forest management can be eradicated by estate taxes that force quick liquidation of land or trees. Stewardship requires legal easements or other conservation mechanisms that will insure the perpetuation of sound forest management and the location of future development where it will not have a significant impact on woodland values.

Most landowners cherish the privacy and solitude afforded by their woods and are deeply inspired by their sense of ownership. These are positive and important values, surely reasons for the purchase and management of woodland, and probably for the purchase of this book. But the true steward is the owner who is always aware of the increasing rarity of a place so beautiful and productive as New England's woods; who understands the ecological meaninglessness of legal boundaries and the short span of a human life relative to that of a forest; and who knows the limits of human understanding in a system so diverse and complex. The true land steward feels a sense of ownership tempered by that knowledge and, even when alone in the woods, walks in the company of people in distant places and generations yet unborn.

Forestry Measurements and Conversions

Map and Deed Description Measurements

Linear:

1 link = 7.92 inches	1 rod = 16.5 feet
100 links = 1 chain	4 rods = 1 chain
1 chain = 66 feet	1 mile = 5,280 feet
80 chains = 1 mile	

24,000:1 is equivalent to 1 inch = 2,000 feet

25,000:1 is equivalent to 1 inch = 2,083 feet

62,500:1 is approximately equivalent to 1 inch = 1 mile

Square:

1 square rod = 272.25 square feet; in old deeds sometimes called a rod

10 square chains = 1 acre

1 acre = 43,560 square feet; a perfectly square acre measures about 209 feet on each side

1 square mile = 640 acres

Tree Size, Volume, and Weight Measurements

Diameter to Circumference Conversion (in inches)

Diameter	6	8	10	12	14	16	18	20	22	24
Circumference	18⅞	25⅛	31⅜	37¾	44	50¼	56½	62⅞	69⅛	75⅜

1 board foot = 144 cubic inches of wood, most often thought of as a piece of wood 1 foot × 1 foot × 1 inch

1 cubic foot = 1,728 cubic inches of wood, most often thought of as a piece of wood 1 foot × 1 foot × 1 foot

1 cunit = 100 cubic feet of solid wood

1 cord = a closely stacked pile of wood containing 128 cubic feet of wood and air spaces, and usually measured as 4 feet × 4 feet × 8 feet; it contains between 70 and 90 cubic feet of wood

1 cubic foot = yields approximately 5 to 6 board feet because of losses in sawing

1 cord = approximately 500 board feet

1,000 board feet of green hardwood logs weighs 5,000 to 9,000 pounds
1,000 board feet of green softwood logs weighs between 4,000 and 6,000 pounds
1 cord of green hardwood weighs 4,000 to 6,000 pounds

The tables below show average volume for a mixture of species, and provide a good approximation of volume in mixed stands. Volume tables that are species specific are more precise, but are not always available or practical to use.

Table A.1. Average Number of Board Feet in 16-Foot Logs, According to the International ¼-Inch Rule

| | Number of 16-Foot Logs in Tree | | | | | | | |
	½	1	1½	2	2½	3	3½	4
DBH (inches)				contents in board feet				
12	30	60	80	100	120			
14	40	80	110	140	160	180		
16	60	100	150	180	210	250	280	310
18	70	140	190	240	280	320	360	400
20	90	170	240	300	350	400	450	500
22	110	210	290	360	430	490	560	610
24	130	250	350	430	510	590	660	740
26	160	300	410	510	600	700	790	880
28	190	350	480	600	700	810	920	1,020
30	220	410	550	690	810	930	1,060	1,180
32	260	470	640	790	940	1,080	1,220	1,360
34	290	530	730	900	1,060	1,220	1,380	1,540
36	330	600	820	1,010	1,200	1,380	1,560	1,740
38	370	670	910	1,130	1,340	1,540	1,740	1,940
40	420	740	1,010	1,250	1,480	1,700	1,920	2,160
42	460	820	1,100	1,360	1,610	1,870	2,120	2,360

Table A.2. Average Number of Cords in Trees of Various Sizes

| | \| Number of 8-Foot Sections in Tree | | | | | | | |
DBH (inches)	1	2	3	4	5	6	7	8
				cords				
4	0.007	0.011						
6	.017	.028	0.040	0.047				
8	.031	.050	.068	.087	0.106	0.116		
10	.049	.082	.111	.133	.160	.188	0.211	
12	.070	.121	.165	.198	.225	.260	.300	0.330
14	.095	.167	.228	.273	.311	.353	.40	.47
16	.122	.220	.300	.367	.42	.47	.53	.59
18	.155	.282	.382	.47	.55	.60	.65	.73
20	.194	.353	.48	.59	.68	.76	.81	.89
22	.240	.44	.60	.73	.84	.93	1.00	1.07
24	.288	.52	.72	.88	1.00	1.12	1.21	1.28
26	.340	.62	.84	1.04	1.19	1.33	1.44	1.51
28	.388	.72	.97	1.20	1.38	1.55	1.67	1.76
30	.43	.80	1.10	1.37	1.59	1.7	1.93	2.04

Landowner's Reading List

An updated version of this appendix is available at
www.upne.com/appendices/woodland.html

Recommended leaflets, pamphlets, and books are listed by chapter, with titles in alphabetical order. Within listings for chapters 2, 5, and 6 reading material is further organized by subject areas. Free publications have an asterisk (*) after the listing, while those costing less than $5 have a dagger (†). Prices should be checked before ordering, since they are subject to change. Periodicals that feature forestry-related articles are listed at the end of this appendix.

LEAFLETS, PAMPHLETS, AND BOOKS

A New England Forest History: Chapter 1

Changes in the Land, by W. Cronon. 1983. Hill and Wang, 19 Union Square West, New York 10003.

The Nature of Vermont, by C. W. Johnson. 1980. University Press of New England, 23 So. Main St., Hanover, NH 03755-2048.

Wildlands and Woodlots, by L. C. Irland. 1982. University Press of New England, Hanover, NH 03755.

Written in Stone, by C. Raymo and M. Raymo. 1989. Globe Pequot Press, 138 W. Main St., Chester, CT 06412.

Assessing Woodland Potential: Chapter 2

BOUNDARIES AND MAPPING
Elementary Forest Surveying and Mapping, by R. L. Wilson. 1985. Oregon State University Bookstores, Inc., Corvallis, OR 97330.

Woodland Boundaries, by H. P. Wood and R. W. Kulis. 1980. Cooperative Extension Service, University of Massachusetts, Bulletin Distribution Center, Cottage A, Thatcher Way, Amherst, MA 01003.*

INSECTS AND DISEASES
Decay Losses in Woodlots, by T. C. Weidensaul, C. Leben, and C. W. Ellett. 1977. Bulletin No. 629. Cooperative Extension Service, Ohio State University Extension Office Information, 2120 Fyffe Rd., Columbus, OH 43210.*

Diseases of Trees and Shrubs, by W. A. Sinclair, H. H. Lyon, and W. T.

Johnson. 1988. Cornell University Press, 124 Roberts Place, Ithaca, NY 14851.

A Guide to Common Insects and Diseases of Forest Trees in the North-eastern United States. 1979. U.S.D.A. Forest Service, Northeastern Experiment Station, P.O. Box 67875, Radnor, PA 19087.*

Insects that Feed on Trees and Shrubs, by W. T. Johnson and H. H. Lyon. 1988. Cornell University Press, 124 Roberts Place, Ithaca, NY 14851.

A Photo Guide to the Patterns of Discoloration and Decay in Living Northern Hardwood Trees, by A. L. Shigo and E. V. Larson. 1969. U.S.D.A. Forest Service Research Paper NE-127. U.S.D.A. Forest Service, North-eastern Forest Experiment Station, P.O. Box 67875, Radnor, PA 19807.†

TREE IDENTIFICATION AND ECOLOGY

Animal Tracking and Behavior, by D. and L. Stokes. 1986. Little Brown and Co. 34 Beacon St., Boston, MA 02108.

Eastern Forests, by J. C. Kricher and G. Morrison. 1988. Houghton Mifflin Co., One Beacon St., Boston, MA 02108.

Field Guide to Trees and Shrubs, by George Petrides. 1972. Houghton Mifflin Co., One Beacon St., Boston, MA 02198.

Know Your Trees, by J. A. Cope and F. E. Winch. 1979. Cornell University, Distribution Center, 7 Research Park, Ithaca, NY 14850.†

North American Trees, by R. J. Preston, Jr. 1976. Iowa State University Press, 2121 So. State Ave., Ames, IA 50010.

The Shrub Identification Book, by G. W. Symonds. 1973. William Morrow and Co., 105 Madison Ave., New York, NY 10016.

Silvics of North American Trees, ed. by R. M. Burns and B. H. Honkala. 1991. U.S.D.A. Forest Service Agricultural Handbook No. 654. Superintendent of Documents, U.S. Government Printing Office, Washington, D.C. 20402.

The Tree Identification Book, by G. W. Symonds. 1977. William Morrow and Co., 105 Madison Avenue, New York, NY 10016.

Why Trees Grow Where They Do in New Hampshire Forests, by W. Leak and J. Riddle. 1979. Publication no. NE-INF-37-79. U.S.D.A. Forest Service, Northeastern Forest Experiment Station, P.O. Box 67875, Radnor, PA 19087.*

Winter Keys to the Woody Plants of Maine, by C. S. Campbell, F. Hyland, and M. L. F. Campbell. 1975. University of Maine Bookstore, Memorial Union, University of Maine, Orono, ME 04469.

Foresters: Chapter 3

Selecting Consulting Foresters: How to Select One and What They Do, by P. Harou et al. 1981. Cooperative Extension Service, University of Massachusetts, Bulletin Distribution Center, Thatcher Way, Amherst, MA 01003.*

Management Plans: Chapter 4

Commonly Used Woodland Management Terms. 1979. Cornell University, Distribution Center, 7 Research Park, Ithaca, NY 14850.*

Forest Management by Compartments, by R. R. Morrow. 1976. Conservation Circular, vol. 14, Nos. 2, 3, and 4. Cornell University, Distribution Center, 7 Research Park, Ithaca, NY 14850.*

Woodland Management Techniques: Chapter 5

FUELWOOD

Fuelwood Production. 1987. Cooperative Extension Service, Iowa State University, Ames, IA 50011.*

Managing Young Stands for Firewood, by K. F. Lancaster. 1980. Publication no. NA-FR-19. U.S.D.A. Forest Service, Northeastern Area State and Private Forestry, P.O. Box 67875, Radnor, PA 19087.*

Woodlot Management for Fuelwood, by M. Koelling and D. Dickmann. 1981. Michigan State University, Rm. 126 Natural Resources Building, E. Lansing, MI 48824.

MAPLE SAP

Amateur Sugar Maker, by Noel Perrin, 20th anniv. ed., 1992. University Press of New England, 23 So. Main St., Hanover, NH 03755-2048.

A Guide to Sugarbush Stocking, by H. D. Smith and C. B. Gibbs. 1970. U.S.D.A. Forest Service Research Paper NE-171. U.S.D.A. Forest Service, Northeastern Forest Experiment Station, P.O. Box 67875, Radnor, PA 19087.*

Maple Sirup Producers Manual, by C. Willits and C. Hills. 1976. U.S.D.A. handbook 134. U.S. Government Printing Office, Washington, DC 20402.†

A Silvicultural Guide for Developing a Sugarbush, by K. F. Lancaster et al. 1974. U.S.D.A. Forest Service Research Paper NE-286. U.S.D.A. Forest Service, Northeastern Forest Experiment Station, P.O. Box 67875, Radnor, PA 19087.†

Sugar Bush Management, by R. R. Morrow. 1976. Information Bulletin 110. Cornell University, Distribution Center, 7 Research Park, Ithaca, NY 14850.†

Sugarbush Management: A Guide to Maintaining Tree Health, by D. Houston, D. Allen, and D. Lachance. 1990. U.S.D.A. Forest Service Technical Report NE-129. Northeastern Forest Experiment Station, P. O. Box 6775, Radnor, PA 19087.

Sugarbush Management for Maple Syrup Producers, by C. F. Coons. 1975.

Division of Forests, Forest Management Branch, Ministry of Natural Resources, Box 434, R.R. 6, Ottawa, Ontario, Canada K1G 3N4.†

RECREATION AND AESTHETICS
Intangibles, by T. D. Rader. 1978. Pennsylvania Forest Resourcs, No. 51. Cooperative Extension Service, Pennsylvania State University, P.O. Box 6,000, University Park, PA 16802.

Trail Building and Maintenance, by R. D. Proudman and R. Rajala. 1981. Appalachian Mountain Club Books, 5 Joy St., Boston, MA 02108.

Using Your Land for Pleasant Living, by B. Wilkins. 1972. Extension bulletin No. 1,211. Cornell University Distribution Center, 7 Research Park, Ithaca, NY 14850.

TIMBER PRODUCTION
Most publications on pruning issued before 1980 are outdated by new research and should not be used.

Choices in Silviculture for American Forests. 1981. Society of American Foresters, 5400 Grosvenor Lane, Washington, D.C. 20014.

Essentials of Forestry Practice, by C. Stoddard, 3d. ed. 1987. John Wiley and Sons, 605 3rd Ave., New York, NY 10016.

An Introduction to Forestry, by G. Sharpe, C. Hendee, and S. Allen. 4th ed. 1986. McGraw-Hill Book Co., Princeton Rd., Hightstown, NJ 08520.

Managing Small Woodlands for Timber Production, by G. Goff and J. Lassoie. No date. Cornell University, Distribution Center, 7 Research Park, Ithaca, NY 14850.*

The Practice of Silviculture, by D. Smith. 1986. John Wiley and Sons, 605 Third Ave., New York, NY 10016.

Proper Pruning, by A. Shigo. 1981. U.S.D.A. Northeastern Forest Experiment Station, Box 64D, Durham, NH 03824.*

The Trees Around Us, by the Nova Scotia Department of Lands and Forests. 1980. Nova Scotia Government Bookstore, P.O. Box 637, 1597 Hollis St., Halifax, Nova Scotia, Canada B3J 2T3.

Woodland Ecology: Environmental Forestry for the Small Landowner, by L. S. Minckler. 1980. Syracuse University Press, 1011 E. Water St., Syracuse, NY 13210.

The Woodland Steward, by J. R. Fazio. 1985. The Woodland Press, Box 3524 University Station, Moscow, ID 83843.

Woodlands for Profit and Pleasure, by R. D. Forbes. 1971. American Forestry Association, 1319 18th Street, N.W., Washington, D.C. 20036.†

TREE PLANTING
Collecting and Planting Seeds of Cone-Bearing Trees, by W. H. Brener and G. R. Cunningham. 1966. University of Wisconsin. Extension Service

Publication no. 535. Cooperative Extension Service, University of Wisconsin, Madison, WI 53706.*

Dollars and Decisions in Tree Planting, by B. Adams. 1972. Publication no. PA-992. U.S. Government Printing Office, Washington, DC 20402.†

Forest Management Chemicals, by D. Hamel. 1981. Agricultural handbook no. 585. U.S. Government Printing Office, Washington, DC 20402.

Forest Tree Planting Guide, by J. H. Noyes. 1980, revised. Cooperative Extension Service, University of Massachusetts, Bulletin Distribution Center, Thatcher Way, Amherst, MA 01003.*

New England Brush Control in Nonfood Crop Areas and Weed and Brush Control in Christmas Tree Stands. May be obtained from local extension agents.*

Southern New England Christmas Tree Growers' Handbook, ed. by Mark Brand. 1992. Cooperative Extension Service, University of Massachusetts, Bulletin Distribution Center, Thatcher Way, Amherst, MA 01003.†

WILDLIFE HABITAT IMPROVEMENT

Building a Pond. Farmers' bulletin no. 2,256. U.S. Government Printing Office, Washington, DC 20402.†

Care of Wild Apple Trees. The Extension Service, University of Vermont, Morrill Hall, Burlington, VT 05402.*

Earth Ponds: The Country Pond Maker's Guide, by T. Matson. 1991. Countryman Press, Woodstock, VT 05091.

How to Attract Cavity-Nesting Birds to Your Woodlot. No date. U.S.D.A Forest Service, Northeastern Area State and Private Forestry, P.O. Box 67875, Radnor, PA 19087.*

Improve Your Forested Lands for Ruffed Grouse, by G. W. Gullion. 1972. Ruffed Grouse Society of North America, 103 N. Price Street, Kingwood, WV 26537.†

New England Wildlife: Habitat, Natural History and Distribution, by R. M. DeGraaf and D. B. Rudis. 1986. U.S.D.A. Forest Service General Technical Report NE-108, Northeastern Forest Experiment Station, P.O. Box 76775, Radnor, PA 19087.*

New England Wildlife of Forested Habitats, by R. M. DeGraaf, M. Yamasaki, et al. 1992. U.S.D.A. Forest Service General Technical Report NE-144, Northeastern Forest Experiment Station, P.O. Box 6775, Radnor, PA 19087.

Managing Small Woodlands for Wildlife, by R. J. Gutierrez et al. 1979. Cornell University, Distribution Center, 7 Research Park, Ithaca, NY 14850.†

Pennsylvania Forest Resources, by T. D. Rader. 1973–1980. Continuing series. No. 6, *The Woodcock;* No. 15, *Winter Birds;* No. 21, *Timber Rat-*

tlesnake; No. 27, *Wild Turkey*; No. 31, *Red Fox*; No. 33, *Ruffed Grouse*; No. 37, *Woodland Salamanders*; No. 50, *Deer Management*; No. 64, *Eastern Coyote*; No. 68, *Woodlands and Wildlife*; No. 69, *Porcupines*; No. 71, *Songbird Management*. Cooperative Extension Service, Pennsylvania State University, P.O. Box 6,000, University Park, PA 16802.

Shrubs and Vines for Northeastern Wildlife, by J. Gill and W. Healy. 1974. General Technical Report NE-9. U.S. Department of Commerce. National Technical Information Service, 5285 Port Royal Rd., Springfield, VA 22151.

Trees, Shrubs, and Vines for Attracting Birds: A Manual for the Northeast, by R. DeGraaf and G. Witman. 1979. University of Massachusetts Press at Amherst, Amherst, MA 01003.

Wildlife Habitat Management for Vermont Woodlands: A Landowner's Guide, by the Vermont Fish and Game Department. 1979. State of Vermont, Fish and Game Department, Montpelier, VT 05602.*

Wildlife and Timber from Private Lands: A Landowner's Guide to Planning, by D. J. Decker, J. W. Kelley, et al. 1990. Cornell University. Distribution Center, 7 Research Park, Ithaca, NY 14850.

Harvesting Forest Products: Chapter 6

FELLING AND SKIDDING TREES
Barnacle Parp's New Chain Saw Guide, by W. W. Hall. 1985. Rodale Press, Emmaus, PA 18049.

Handbook for Eastern Timber Harvesting, by F. C. Simmons. 1979. U.S.D.A. Forest Service. Stock no. 001-001-00443-0. U.S. Government Printing Office, Washington, DC 20402.

Skidding Firewood with Small Tractors, by D. Stevens and N. Smith. 1980. Publication no. NA-GR-10. U.S.D.A. Forest Service Northeastern Area State and Private Forestry, P.O. Box 67875, Radnor, PA 19087.*

Tree Felling, by T. McEvoy. 1979. Cooperative Extension Service, University of Connecticut, Storrs, CT 06268.†

LOGGING ROADS AND EROSION CONTROL
Permanent Logging Roads for Better Woodlot Management, by Richard F. Haussman and E. W. Pruett. 1978. U.S.D.A. Forest Service, Northeastern Area State and Private Forestry, P.O. Box 67875, Radnor, PA 19087.*

Acceptable Management Practices for Maintaining Water Quality on Logging Jobs in Vermont, 1987. Vermont Department of Forest Parks and Recreation, 103 Main St., Bldg. 10S, Waterbury, VT 05676.*

Erosion and Sediment Control Handbook for Maine Timber Harvesting Operations: Best Management Practices. 1991. Maine Forest Service, Station #22, Augusta, ME 04333.*

Massachusetts Best Management Practices, Timber Harvesting Water Quality Handbook, by David Kittredge, Jr. and Michael L. Parker. 1989. Cooperative Extension Service, University of Massachusetts, Bulletin Distribution Center, Thatcher Way, Amherst, MA 01003.†

SELLING TIMBER AND FUELWOOD

Harvesting and Marketing Your Timber in Connecticut, by T. J. McEvoy. No date. Cooperative Extension Service, University of Connecticuit, Storrs, CT 06268.†

How to Estimate the Value of Timber in Your Woodlot, by H. V. Wiant, Jr. 1989. Circular 148, Agricultural and Forestry Experiment Station, West Virginia University, Morgantown, WV 26506.

Marketing Timber from the Private Woodland, by R. B. Heiligmann and M. R. Koelling. 1979. Cooperative Extension Service, Michigan State University, East Lansing, MI 48824.*

When to Harvest Your Trees, by K. D. Coder and D. W. Countryman. 1979. Cooperative Extension Service, Iowa State University, Ames, IA 50011.*

Financial Aspects of Forest Management: Chapter 7

Charitable Gifts of Land: A Landowner's Guide to Vermont and Federal Tax Incentives, by Darby Bradley. 1982. Vermont Land Trust, 8 Bailey Ave., Montpelier, VT 05602.†

Financial Aspects and Investment Potential of Timber Management in Vermont, by M. Beattie. 1982. The Extension Service, University of Vermont, PubBR 1-338, Publications Department, Morrill Hall, Burlington, VT 05405.*

Forest Owners' Guide to Timber Investments: The Federal Income Tax and Tax Recordkeeping, by W. Hoover, W. Siegel, G. Myles, and H. Hanley, Jr. 1989. U.S.D.A. Forest Service Agriculture Handbook No. 681. Supt. of Documents, U.S. Government Printing Office, Washington, DC 20402.

Tax Savings on Timber Sales, by J. E. Bethune. 1981. Connecticut Cooperative Extension Service, University of Connecticut, College of Agriculture and Natural Resources, Storrs, CT 06268.*

What Forest Landowners Should Know About Federal Estate and Gift Taxes, by K. Utz, W. Siegel, and J. Gunther. 1978. General Report SA-GRI, U.S. Government Printing Office, Washington, DC 20402.*

Woodland Account Book: A Basic Record Keeping System for the Private Woodland Owner, by M. Patmos. No date. The Cooperative Extension Service, University of New Hampshire, Durham, NH 03824.

PERIODICALS

American Forests. American Forestry Association, P.O. Box 2000, Washington, DC 20013–2000. Bimonthly.

Tree Farmer. American Forest Foundation, 1111 19th Street N.W., Washington, DC 20036. Bimonthly.

Forestry Laws

An updated version of this appendix is available at
www.upne.com/appendices/woodland.html

This appendix explains some of the more important regulations in each New England state that pertain to forest lands. State by state, laws are listed under one or more of the following alphabetical categories:

> Boundaries
> Current Use Assessment
> Cutting Practices
> Liability of Landowners
> Logging in Fragile or Specially Designated Areas
> Open-Air Fires
> Petition for Right-of-Way
> Registration of Forestry-Related Workers, Equipment, or Businesses
> Slash Disposal
> Threatened and Endangered Species
> Timber Trespass
> Weights and Measures
> Yield Tax

Specific laws are identified by number according to the generally accepted format for each state.

"Current use assessment" or "use value" programs allow the calculation of property taxes for eligible forest land to be based on some standard other than fair market value. In most New England states, the alternative value used is that of the merchantable timber that an acre of forestland can produce in a year. This alternative value usually results in a much lower property assessment than does fair market value. The purpose of these programs, which are legislatively enacted, is usually to reduce the tax pressure on an owner that leads to the subdivision and development of productive forest land. Each New England state has a current use assessment program, but application procedures to be followed by landowners and eligibility requirements for forest land vary, as shown in this appendix.

It should be noted that this appendix is based on the authors' understanding of the various state regulations and should be used only as a general reference. The appendix does not include all laws pertaining to forest land. Because laws and their legal interpretations are subject to change, the state, county, or extension foresters listed in appendix 4 should be consulted for current and detailed information. There are federal laws that take precedence in certain circumstances over some of the state laws listed here, notably endangered species and wetlands laws. Public agencies listed in appendix 4 can provide information on these federal laws.

CONNECTICUT

Current Use Assessment

SECTION 12 – 107

Eligibility Requirements:

(1) A landowner must own a total of 25 acres of forest land, with each noncontiguous tract being at least 10 acres in size.

(2) Forest land not only includes land with tree growth, but also ledges, streams, access roads, shallow water, open marshes, power lines, and abandoned fields with 500 trees per acre. Plantations are eligible for current use assessment after one successful growing season.

(3) A map of the property must be included in the application. It must be in a specified format, must show property boundaries and delineate forested areas, and include an estimate of total and forested acreage.

(4) A town road map showing the location of the property must be included in the application.

(5) Applications are due by October 31st. However, in a year when revaluations become effective, an additional 60 days are permitted.

Ongoing requirements:

Land must be maintained as forest land. No management is necessary.

Recommended use values:

The state has one suggested maximum use value for forest land in river towns, and another for the rest of the state.

Withdrawal:

Land is withdrawn from classification if the property is sold or its use has changed to a use other than farm, forest, or open space. A conveyance tax is owed by the person selling the land or making the land-use change. The tax is based upon the sale price of the land, or, in the case of land-use change, its fair market value. The tax percentage is based on the number of years during which the land has been classified or owned, whichever is greater. If the land has been classified or owned for one year, the tax is 10 percent of the sale price of fair market value; taxation is reduced by one percentage point for each subsequent year up to the tenth year, when no conveyance tax is due.

Cutting Practices

PA 91–335, SECTIONS 23–65P

(1) Foresters and timber harvesters are required to be certified.

(2) The Commissioner of the Department of Environmental Protection (DEP) is empowered and directed to regulate the conduct of forest practices in the state.

(3) Municipalities are allowed to regulate forest practices, provided their ordinances or regulations have been approved by the Commissioner of DEP as conforming to the intent of the law.

(4) A Forest Practices Advisory Board advises the Commissioner in the certification of forest practitioners, the regulation of forest practices and the forest-based programs and policies of the DEP.

Liability of Landowners

SECTION 52–557i

No landowner shall be liable for injury sustained by persons operating or passengers of snowmobiles, all terrain vehicles, motorcycles, minibikes, or minicycles unless a fee has been charged or injuries are a result of willful or reckless conduct by the owner.

SECTION 52–557k

If land is used for the cutting of firewood, from standing trees or slash, the owner will not be liable for injuries whether a fee is charged or not, unless the owner sells more than 100 cords of wood each year, or the injuries are a result of the owner's failure to warn of any hidden hazard which is known to the owner.

Logging in Fragile or Specially Designated Areas

SECTION 22a–40

A permit is needed if an owner proposes to disturb the natural indigenous character of a wetland or waterway by the removal or deposition of material, alteration or obstruction of water flow, or pollution of wetland or waterway. A plan must be submitted to local inland wetland commissions or, if the town does not have one, the State Bureau of Water Resources. Farming operations do not need permits. *According to Section 1–1, the term "farming" includes forestry practices; however, it is recommended that before disturbing wetlands or water courses an inland wetland commission be consulted.*

Some towns in Connecticut have local ordinances regulating harvesting operations where wetlands or waterways might be disturbed. Town officials should be consulted about local ordinances. Town regulations must conform to the intent of the state forest practices act.

Open-Air Fires

SECTION 19 – 508

A permit is required to burn brush in open-air fires. No permits will be granted if air quality standards may not be met, or when there is a danger of fire within 100 feet of woodland, or grassland.

Slash Disposal

SECTION 23 – 46

Brush shall not be piled within 25 feet of a highway, or within 100 feet of any building for more than ten days.

Threatened and Endangered Species

TITLE 26, SECTIONS 303 – 315

It is illegal: (1) for any person to take any threatened or endangered wildlife or plant species from public property or the waters of the state; (2) for any person to take endangered species from private property without the written permission of the landowner; and (3) for private landowners to sell or transport threatened or endangered species from their property. Takings by private landowners that are incidental to legal activities on their own land are exempt. Species listing and identification of essential habitats are the responsibility of the Commissioner of the Department of Environmental Protection.

Timber Trespass

SECTION 52 – 560

Persons who without permission willfully cut down, destroy, or carry away trees, timber, or wood from the land of another shall be liable for three times the value of the same, except in the case of Christmas trees, for which liability is five times the value. If persons acted by mistake or had good reason to believe the land was theirs, liability is limited to actual (single) value.

Weights and Measures

SECTION 43 – 26

The international log rule is the standard log rule for determining the board foot content of sawlogs, and all contracts shall be on the basis of this rule unless some other measure is agreed to.

SECTION 43 – 27

All fuelwood shall be sold by the standard cord or a fraction thereof, or by weight. A cord contains 128 cubic feet of compactly piled wood. The terms "facecord," "rack," "pile," and "truckload" cannot be used when advertising or selling fuelwood. A dealer of fuelwood who sells more than 25 cords per year may not sell fuelwood by weight or load unless the wood is weighed by a public weigher. The sale of fuelwood requires a ticket stating information concerning weight of wood, names and addresses of all parties (including the weigher), price, and whether the wood is green or seasoned.

MAINE

Boundaries

TITLE 14, SECTION 7554

Any person removing or destroying any lawfully established survey markings along a boundary line shall be liable for the cost of the damage done, including necessary engineering services.

Current Use Assessment

TREE GROWTH TAX LAW
TITLE 36, SECTIONS 571–584A

Eligibility requirements:

(1) A parcel must be greater than 10 acres of forest land. Forest land does not include swamps, marshes, ledges, and similar areas which are unsuitable for growing trees, even though such an area may exist within forest land.

(a) A landowner must have a written forest management plan which is being followed.

(b) A sworn statement is needed that the owner is managing the land according to accepted forestry practices with the objective of growing trees of commercial value.

(2) A map of the property must delineate forest types and list the number of acres in each type.

(3) Applications are due April 1st at the local assessors' office.

Ongoing requirements:

A landowner must follow the plan and must submit, every 10 years, a statement from a licensed forester that the landowner is managing the parcel according to the schedules in that plan.

Recommended use values:

The state tax assessor determines values for each forest type; these are later adjusted by the percentage of current value (the equalization ratio) that is being applied to other property in the town. There is a reduction in assessment for land containing a low density of trees.

Withdrawal:

When land is withdrawn from classification because it no longer meets the requirements, the owner at the time of withdrawal is subject to the greater of the following taxes:

(1) The taxes that would have been assessed on the land at fair market value for up to five years from the time the land was classified, plus interest charges on that amount for those years. The taxes which have been paid during the period under the Tree Growth Tax Law, plus interest on those taxes, are credited against the amount owed.

(2) The difference between the land's fair market value for the previous year and 100 percent of its current use value for the previous year, multiplied by one of the following factors: 30 percent for property withdrawn after one year in the program and 1% lower for each additional year until the 20% floor is reached.

Forest land does not become declassified when it is sold, but when its use changes. When land transfers ownership the new owner has one year to have a new management plan written.

Cutting Practices Report

TITLE 12, SECTION 805

Prior to commencing harvesting operations, the landowner needs to notify the Forest Bureau of any harvest operation and include a map locating the harvest site. Forest land owners, when selling stumpage, are required to report the following to the Forestry Bureau during the month of January of the year after the sale: species, volume, stumpage price, type of harvest, whether it was for land conversion purposes, and location of the harvest. Owners who harvest trees for their businesses must report all but the stumpage price. Precommercial practices on 10 acres or more must also be reported.

Regeneration harvests must meet criteria developed by the Bureau of Forestry which include residual basal area/acre guidelines. Clearcuts in category 1 are greater than 5 acres and less than 35 acres. The separation zone between clearcuts is 250 feet. If it is greater, it will be considered a single clearcut and if the total acreage is greater than 35 acres it is a category 2 clearcut. A category 2 clearcut is greater than 35 acres and less than 125 acres in size. The separation zone between these clearcuts is at least 1.5 times the total area contained within the perimeter of the clearcut. Harvesting in the separation zone is regulated. All category 2 clearcuts must file a certification that the regeneration standards have been met. Larger acreage must follow stricter guidelines. Management plans are required on all clearcuts over 50 acres with provisions of how the regeneration standards will be met.

Liability of Landowners

TITLE 14, SECTION 159A

If land is used for recreational purposes or for harvesting forest products, and no fee is charged and no compensation received, the owner shall not be liable for any damages unless injuries are a result of willful or reckless conduct by the owner. An owner is responsible if a person who is paying a fee for the use of the land is injured by a person harvesting wood.

Logging in Fragile or Specially Designated Areas

LAND USE REGULATION COMMISSION (LURC)
TITLE 12, SECTIONS 681–689

In unorganized towns, road building and timber harvesting require permits in areas known as protection districts. These include winter deer-yards, steep slopes, 250-foot strips along shorelines, and elevations above 2,500 feet. All protection districts are mapped by the LURC office in Augusta.

SHORELAND ZONING
TITLE 12, SECTIONS 4811–14

In organized towns where land is within 250 horizontal feet of desig-nated lakes, ponds, rivers, wetlands, or salt water, timber harvesting is regu-lated. Only a percentage of timber volume may be removed, and logging roads must be located in a specified fashion. Zoning regulations vary from town to town, and in areas where a town has not adopted local zoning, state guidelines are in effect.

TITLE 12, SECTION 519

Within 100-foot strips of numbered highways, or within 75 feet of des-ignated highways, the harvesting of forest products is limited to the re-moval of 40 percent of the tree cover unless trees are dead or dying, a road is to be built, or the state forester gives approval.

TITLE 12, SECTIONS 7754 AND 7755

Any project which is wholly or partly within an "essential habitat" of vertebrate species of wildlife and is permitted, licensed, funded or carried out by a state agency or municipal government, requires approval from the Department of Inland Fisheries and Wildlife. In general, a permit will be granted for forest management activities within the essential habitat of bald eagles if it occurs during the non-nesting period of this endangered species.

Open-Air Fires

TITLE 12, SECTION 1401

Permits are required for burning land, for clearing and for burning brush, slash, dry grass, or blueberry land, except when the ground is cov-ered with snow. Permits are obtained from forest-fire rangers in unorga-nized towns, and from forest-fire wardens in organized townships.

Registration of Forestry-Related Workers, Equipment, or Businesses

TITLE 32, SECTIONS 5001–19

Anyone who engages in work as a forester must be registered unless work-ing on their own land or on land held by contractual agreement. Four years of forestry school are required plus two years of acceptable forestry experience. However, two years of acceptable experience may be substituted for each school year missed including a two-year internship. In addition, the successful completion of an examination is required for all foresters.

Slash Disposal

TITLE 38, SECTION 417

No slash resulting from logging operations is permitted on tidal land or in tidal waters, or on banks where it can be washed into those waters.

TITLE 12, SECTIONS 1552–57

No slash resulting from logging operations may remain within 50 feet of a right-of-way of any public highway.

No slash may remain within 25 feet of a boundary if it is declared a fire hazard.

Threatened and Endangered Species

TITLE 5, SECTION 3315

Threatened and endangered plants are listed for informational purposes only.

(For Maine law relating to rare and endangered wildlife see Title 12, Section 7754 and 7755 under Logging in Fragile or Specially Designated Areas, above.)

Timber Trespass

TITLE 14, SECTION 7552

Any person who cuts down, destroys, or carries away the timber, wood, or property of another, if willfully done without license of the owner, is liable for treble damages plus any expenses necessary for the determination of damages and for attorney and court fees.

TITLE 14, SECTION 7552A

When 10 acres or more are to be harvested, and the cutting is to take place within 200 feet of the property boundaries, the owner authorizing the harvest is responsible for clearly marking the property line. If a trespass occurs, an owner may be subject to double damages if boundaries have not been marked.

Weights and Measures

TITLE 10, SECTIONS 2302–2628

A "cord of wood" is defined as 128 cubic feet, and is generally thought of as 4 feet wide, 4 feet high, and 8 feet long when well stacked. Fuelwood when sold loose should be sold by the cubic foot, unless other arrangements are made between buyer and seller.

MASSACHUSETTS

Boundaries

CHAPTER 266, SECTION 94

No person may injure or remove a monument or tree which designates the boundaries of property.

CHAPTER 266, SECTION 105

No person may pull down stone walls or fences which are erected for the purpose of enclosing land.

Current Use Assessment

CHAPTER 61

Eligibility requirements:

(1) A landowner must own at least 10 acres of contiguous forest land, exclusive of a house site. Forest land does not include swamps, open water, or land cleared for power lines. Owners of more than 10 acres need not put any additional holdings in the program.

(2) A landowner must have and follow a ten-year management plan that has been certified by the Department of Environmental Management (DEM). The goal of the plan must be to improve the quantity and quality of a continuous forest crop.

(3) A landowner must pay an 8 percent stumpage tax on all forest products harvested during the two years prior to certification.

(4) Property boundaries must be clearly marked on the ground.

(5) Applications and completed management plans are due in the regional DEM office by July 1 of the year preceding that for which certification is sought.

(6) Certificates approved by DEM must be submitted to the Board of Assessors of the town in which the land is located by September 1 of the year preceding that for which certification is sought.

Ongoing requirements:

(1) A certified management plan must be filed every ten years.

(2) A landowner must pay an 8 percent stumpage tax to the town for all forest products harvested. The tax is based on the stumpage value of the trees—that is, their value before they are cut.

Recommended use values:

Land is assessed for use value at 5 percent of its fair market value, or at $10 per acre, whichever is greater.

Withdrawal:

When forest land is withdrawn from classification because it no longer meets the requirements, the owner at the time of withdrawal shall pay a penalty tax. The tax is the difference between taxes paid under Chapter 61 and what would have been paid under full assessment for the period since last certification, or five years, whichever is greater. Interest is charged on the amount due. If land is voluntarily withdrawn at the end of a ten-year management period, a credit will be given for the 8 percent products tax paid.

The city or town where the land is located has the right of first refusal— that is, a chance to match any private offer made, or to pay fair market value for the land if a change of use is planned.

Forest land will not be declassified if a portion of the land is used for a residence by an immediate relative, or if the land is sold but the new owner continues to follow the forest management plan.

CHAPTER 61A

Under Chapter 61A, agricultural and horticultural land is assessed at its value for that purpose.

Eligibility requirements:

(1) Must be at least 5 acres.

(2) The land must produce annual gross sales of at least $500.00, plus $5.00 for each additional acre, except in the case of woodland and wet-land, for which the added value is $.50 per acre.

(3) Woodland is eligible for classification under Chapter 61A, but DEM must certify that the land is being managed under an approved forest management plan, comparable to what is required for Chapter 61.

(4) Application for classification under Chapter 61A must be made annually by October 1 to the local Board of Assessors.

Recommended use values:

A range of allowable values for each category of agricultural land is determined annually by the Farmland Valuation Advisory Commission.

Withdrawal:

(1) When land is sold and converted from an eligible to an ineligible use, a conveyance tax is assessed to the seller. The tax ranges from 1 percent to 10 percent of the sale price, depending on the number of years the land has been in eligible use.

(2) When land is converted from an eligible to an ineligible use without being sold, a rollback tax is assessed. The rollback tax is the difference between actual property tax paid and what would have been paid under full assessment, for a retroactive period of 5 years. Interest is also assessed.

(3) If a classified parcel changes from eligible farmland use to a residential, industrial, or commercial use, with or without a sale, the municipality has a right of first refusal for 120 days to acquire the land at the offered price or appraised fair market value.

CHAPTER 61B

The purpose of Chapter 61B is to prevent the loss of open space and encourage the use of land for recreational purposes.

(1) Must be at least 5 contiguous acres.

(2) The land must be retained in a natural state, be devoted primarily to recreational use, and be open to the public.

(3) Application must be made to the local Board of Assessors before October 1 of each year.

Recommended use values:

Assessed value may not exceed 25% of the fair market value.

Withdrawal:

Identical to Chapter 61A.

Cutting Practices

CHAPTER 132, SECTIONS 42–44

(1) Owners wishing to cut timber must submit a cutting plan to the regional office of the DEM and to the local conservation commission.

(2) Abutting landowners whose land is within 200 feet of the cutting area must be notified, by certified mail or hand delivery, of the operation.

(3) Owners whose land is classified under Chapter 61 must provide an estimate of the stumpage value on the cutting plan.

(4) Boundaries within 50 feet of the cutting area must be blazed or painted.

(5) Upon approval of the cutting plan, DEM issues a permit, which must be displayed at the job site. If DEM fails to act on a cutting plan within 10 business days, work may proceed according to the cutting plan, except in wetland areas.

(6) DEM may levy a fine of up to $100 per acre when a landowner or stumpage owner fails to file a plan or comply with an approved plan. DEM may also issue a stop order when an operation does not comply with the requirements of the law.

(7) Landowners are exempt from the cutting plan requirement if: an owner or tenant is cutting for his/her own noncommercial use; the volume is less than 25,000 board feet or 50 cords; land is being cleared for a different use; cutting is for pasture maintenance.

Liability of Landowners

CHAPTER 21, SECTION 17C

If land is used for recreational purposes, and no fee is charged, the owner shall not be liable for injuries to the users or their property unless the injuries are a result of willful or reckless conduct by the owner.

Logging in Fragile or Specially Designated Areas

WETLANDS PROTECTION ACT
CHAPTER 131, SECTION 40

The filling, dredging, or altering of any bank, freshwater wetland, coastal wetland, beach, dune, flat march, meadow, or swamp bordering on the ocean, any estuary, creek, river, stream, pond, or lake, or any land subject to tidal action, coastal storm, or flooding shall be described in a written notice to be filed along with plans of the proposed activity. Plans should be sent to the town conservation commission, board of selectmen, or mayor. No work shall begin until a permit is granted. The permit may require modifications to the original plans. Normal maintenance and improvement activities on land in agricultural use are exempt from the law. Forest land under a planned program is considered agricultural. When an approved cutting plan is in operation, forestry operations are exempt from the filing requirements of the Wetlands Protection Act.

Open-Air Fires

CHAPTER 48, SECTION 13

A permit is required from either the fire warden or the chief of the fire department to make an open-air fire.

Rare and Endangered Species

CHAPTER 131A

(1) It is illegal to take, possess, buy or sell any plant or animal listed by the state or federal government as endangered, threatened, or of special concern.

(2) Exceptions are made when: the department of public health certifies that a public health hazard exists; and for the purpose of propagation of an endangered plant species, as long as the source used for propagation is not taken from the wild.

(3) No one may alter "significant habitat" without a permit. Designation of significant habitats takes into account "foreseeable uses of the land" as well as population status.

(4) Location of significant habitats and a list of the record owners is recorded in the appropriate registry of deeds.

(5) An owner of significant habitat can petition the state to purchase the land, if the petition is made within 21 days of habitat designation.

(6) Permits to "alter significant habitat" may be issued following submission of: a full set of project plans; alternatives to the proposed project; plans for protection of endangered species and mitigation measures to be taken; information on the potential economic effects of the proposed project.

(7) Landowners whose forested property has been certified by the Endangered Species Program and the State Forester are exempt from the Alteration Permit process. Certification is based on the management plan's sensitivity to endangered species and its specification of activities to "improve the quantity and quality of a continuous crop."

(8) A landowner aggrieved by state agency decisions may file a superior court action to determine whether the state's actions constitute a taking requiring compensation under the Constitution.

(9) Penalties for violation include fines ranging from $500 to $20,000 and prison sentences up to 6 months.

Registration of Forestry-Related Workers, Equipment, or Businesses

CHAPTER 132, SECTION 46

Anyone engaged in the business of harvesting timber or other forest products is required to have a license from the Department of Environmental Management. The license expires June 30th of each year.

Slash Disposal

CHAPTER 48, SECTIONS 16, 16A

Slash from wood or timber harvests which are subject to the Cutting Practices Act (Chapter 132) must be treated in the following manner:

Hardwood slash may not remain more than 2 feet above the ground within 40 feet of railroads, the outer edge of any highway or woodlands of another owner, or within 20 feet of the outer edge of any forest or woods road.

Softwood slash may not remain on the ground within 40 feet of railroads, the outer edge of highways or woodlands of another owner, and may not lie more than 2 feet above the ground between 40 feet and 100 feet of the outer edge of any highway, or 25 feet from the edge of any forest or woods road.

When cutting is not subject to the Cutting Practices Act (Chapter 132), slash should not remain on the ground within 40 feet of woodlands of another owner, of railroads, or within 100 feet of the center of any highway or the outer edge of a multilaned highway.

All slash should be disposed of in such a manner as to minimize danger from fire.

No slash is permitted within 25 feet of any continuously flowing brook or stream, pond, river, or water supply.

Timber Trespass

CHAPTER 242, SECTION 7

Any persons who without permission willfully cut down, destroy, or carry away trees, timber, or wood from the land of another shall be liable for treble damages. If such persons acted by mistake, or had good reason to believe the land was theirs, only single damages are required.

Yield Tax

A yield tax is applicable only in the context of the current use assessment program, Chapter 61.

Weights and Measures

CHAPTER 96, SECTION 11A

The international quarter-inch log rule is the standard measurement for measuring board-foot content. All contracts for the purpose of the purchase or sale of logs shall assume this rule, unless otherwise stated in the contract.

CHAPTER 94, SECTIONS 298, 299

The law defines the terms "cordwood" to mean wood 4 feet in length; "firewood" to mean wood cut to any length between 8 inches and 4 feet; "kindling" to mean wood averaging 8 inches in length.

Cordwood and firewood must be sold or advertised in terms of cubic feet or cubic meters. The terms "cord," "facecord," "pile," or "truckload" shall not be used. A delivery ticket will be provided by the seller of the cordwood or firewood to the purchaser at the time of delivery. It should include the names and addresses of the seller and the purchaser, the quantity delivered in cubic measurement, the date of delivery, and the price.

NEW HAMPSHIRE

RSA = Revised Statutes Annotated of New Hampshire, chapter: section

Boundaries

RSA 472

If owners of abutting land cannot agree on boundary lines because of the destruction of monuments or other boundary markings, they may hire a surveyor and agree to newly established lines in writing, which will then be recorded in their deeds.

Current Use Assessment

RSA 79A

Eligibility requirements:

Woodland may qualify for classification under one of three categories—forest land, unmanaged forest land, or unproductive land or any combination as long as the eligibility requirements are met. All applications are due by April 15th at the local assessors' office.

(1) Managed forest land:

(a) A parcel must be at least 10 acres of contiguous acreage, or a certified tree farm.

(b) The objective must be the growing and harvesting of repeated forest crops.

(c) The assessing officials may require a five-year management plan.

(2) Unmanaged forest land: A parcel must be at least 10 acres of land which is capable of producing commercial forest crops but the landowner is not following responsible land stewardship practices. The land must not have detrimental structures on it.

(3) Unproductive land: A parcel must be at least 10 acres of unimproved land which is by nature incapable of producing agricultural or

forest crops, and which has no detrimental structures on it. Wetlands have no minimum acreage requirement.

Ongoing requirements:

(1) If the local assessing officials require the landowner to show responsible land stewardship is being followed, at intervals of 5 years the following documentation is required: a statement of past forest accomplishments, present forestry conditions, 5-year harvesting plans, and a map showing forest types and acreages. If a landowner provides documentation that the land is a certified tree farm, only a map is needed.

(2) For unproductive land or unmanaged forest land, the land must be left in a natural state, except for the harvesting of fuelwood for the owner's personal use on unmanaged forest land.

Recommended use values:

Forest land is assessed according to the forest type and the characteristics of the land. An assessment rate for unproductive land, forest land, and unmanaged forest land must fall within a stated range as determined by the state. *In general, the values increase from unproductive land to forest land to unmanaged forest land.* There is a 20 percent reduction in assessment if the land is open to public recreational use without a fee. All use values are adjusted according to the percentage of current value (the equalization ratio) that is being applied to other property in the town.

Withdrawal:

If there is a change in the use of the land, or the parcel when sold does not meet minimum acreage requirements, a Use Change Tax must be paid. A change in use means excavation or grading for future construction or for mining, or the installation of utilities. The tax is 10 percent of the full value of the land.

Cutting Practices Report

RSA 79:10, 10−b

A "notice of intent to cut" must be filed with assessing officials and with the State Commissioner of Revenue Administration. If logging continues after March 31st, another notice is required. Certificates sent by the Commissioner of Revenue Administration, stating that the notice of intent has been filed, should be posted in the area of cutting. If no timber is cut, a report so stating must be filed. After cutting, assessing officials may require a report of all wood and timber cut during the tax year. If more volume is cut than had been estimated, an amended notice of intent to cut is required.

All of these reports must be filed by the owner of the land.

The report is not required from persons who cut timber or wood from their own land for personal use.

Liability of Landowners

RSA 212:34

If land is used for recreational purposes or for the cutting of cordwood, and no fee is charged, the owner shall not be liable unless injuries are a result of willful or reckless conduct by the owner. Landowners may cut up to 10,000 board feet or 20 cords.

Open-Air Fires

RSA 224:27

Burning permits are required for open-air fires except when the ground is snow-covered. Permits can be obtained from the fire wardens.

Petition for Right-of-Way

RSA 234—A

If it is necessary for the convenient removal of lumber or wood to pass through lands of a person other than the owner of the lumber or wood, the selectmen, if petitioned, may lay out a right-of-way. Damages will be assessed and paid before the right-of-way is used. The selectmen can set up conditions for use.

Logging in Fragile or Specially Designated Areas

RSA 149:1 – 19

A permit is needed to change the course of natural runoff, or to create new, unnatural runoff courses for surface waters. Plans must be sent to the Water Supply and Pollution Control Commission.

RSA 224:44 – a

When land is used for timber growing and other forest products, and is within 150 feet of any pond larger than 10 acres, or of any navigable river, stream, brook, or public highway, or is within 50 feet of any stream, river, or brook which flows throughout the year, no more than 50 percent of the basal area may be cut, and the remaining trees must be well distributed, unless written consent is obtained from the state forester.

RSA 482 – A

If a forestry operation is to minimally impact on a wetland (fewer than 3,000 square feet will be disturbed), then the landowner must notify the New Hampshire Wetland Board and the town conservation commission using the notification form attached to the notice of intent to cut. If the forestry will have major impact on a wetland (greater than 3,000 square feet will be disturbed), then the landowner must obtain a dredge and fill permit from the New Hampshire Wetlands Board.

RSA 672:1, III c

Forestry activities including the harvest and transportation of forest products shall not be unreasonably limited by town planning or zoning.

Registration of Forestry-Related Workers, Equipment, or Businesses

RSA 224 – A

Sawmills must be registered annually.

RSA 98 – 117

Any person engaging in the practice of forestry must be licensed by the New Hampshire Joint Board. Licensed foresters must meet minimum education and forest experience criteria. For renewal of license, foresters must participate in continuing education courses.

Slash Disposal

RSA 224:44 – b

Timber, brush, lumber, or wood may not remain in any river, stream, or brook. It may not remain within 25 feet of the land of another owner, or of any river, stream, or brook which will float a canoe; within 50 feet of any pond larger than 10 acres, or of any navigable river, or the nearest edge of a public highway; within 60 feet of a railroad right-of-way; or within 100 feet of any occupied building except a temporary lumber camp. Slash must not extend more than 4 feet above the ground; it must not be left within 50 to 150 feet of any pond larger than 10 acres, or of any navigable river, stream, brook, public highway. A *navigable waterway means a waterway deep enough and wide enough to carry a boat or vessel.*

Threatened and Endangered Species

RSA 212 – A

Threatened and endangered wildlife species are listed at the determination of the Director of the Department of Fish and Game. The Director can conduct research to determine the viability of a species. It is unlawful to take, possess, export or ship a species that has been determined to be rare or endangered.

RSA 217 – A

Threatened or endangered plant species cannot be removed from a property without permission of the landowner.

Timber Trespass

RSA 539:1

Anyone who without permission willfully cuts down, destroys, or carries away trees, timber, or wood from the land of another shall be liable for five times the value of the same.

Weights and Measures

RSA 79:1

The international quarter-inch log rule is the standard measure for determining the value of logs cut for the yield tax. If any other log rule is used, it must be converted to international standard.

RSA 359:A, A35; RSA 339:16

A cord is defined as the amount of wood contained in a space of 128 cubic feet when the wood is closely stacked. Normal shrinkage resulting from cutting wood into shorter lengths will not be interpreted as a short measure. It is only legal to sell wood by the cord or by a fraction of a cord, unless both parties agree otherwise.

Yield Tax

RSA 79:3, 3a

A yield tax should be paid at the rate of 10 percent of the stumpage value at the time of cutting. When open competitive bids have been the basis of a sale, the price paid is the assessed value upon which the tax is based. In other cases the assessors may take into account open bids in the vicinity and operating costs to determine assessed value.

Persons intending to cut trees must file a bond or a security deposit with the assessors if they do not own real estate in the town where trees are to be cut.

RHODE ISLAND

Boundaries

SECTION 2 – 15 – 2

Any persons who intend to cut trees must notify the Department of Environmental Management five days before such cutting. Cutting on single holdings of less than 5 acres, or cuttings of fewer than 5,000 board feet or 25 cords in one year, are exempt.

Current Use Assessment

FARM AND FOREST OPEN SPACE TAX
SECTION 44 – 27 – 1 TO 44 – 27 – 6;
SECTION 44 – 5 – 39 TO 44 – 5 – 41;

Eligibility requirements:

(1) Forest land must total 10 acres, not including a zoned house lot or 1 acre surrounding any dwelling, whichever is smaller. Owners of more acreage need not put the entire holding under classification.

(2) A landowner must have a five-year management plan drawn up by a forester and approved by the Department of Environmental Management (DEM).

(3) Land can be classified as forest land if it is managed for any of the following objectives: timber, Christmas trees, wildlife, watershed, soil stability, or noncommercial recreation.

(4) Applications are due for certification by the DEM by November 1st. However, in a year when revaluation takes place, they are due twenty days after receipt of notice of revaluation or receipt of a tax bill.

Ongoing requirements:

(1) A landowner must annually certify that land is being managed as forest land.

(2) The management plan must be updated every five years.

Recommended use values:

Assessing officials in each town will determine the assessed value based on its current use, without regard to neighboring land values of a more intensive nature.

Withdrawal:

If any owner fails to follow the management plan, or changes the use of the land to one inconsistent with forestry, the land is withdrawn from the program. The land-use change tax is determined at the time the land is withdrawn for any reason. If the land has been classified for up to six years, the tax is 10 percent of the fair market value. The percentage of taxation is reduced by 1 percentage point for each additional year of classification, until no tax is due for the fifteenth year. The payment of the land-use change tax is not due until development has occurred.

Liability of Landowners

SECTION 32 – 6 – 1 TO 32 – 6 – 7

Landowners wishing to limit their liability toward recreational users of their land must apply to the Department of Environmental Management. The department will then inspect the property for dangerous conditions. If the department then accepts the application, the landowner shall not be liable for injuries to recreational users of the property unless the injuries are a result of willful or reckless neglect by the owner, or unless a fee has been charged for use.

Logging in Fragile or Specially Designated Areas

SECTION 2 – 1 – 18

Any person who is planning to disturb a wetland must apply to the Wetlands Division of the Department of Environmental Management for a permit. Wetlands include bogs, swamps, marshes, river banks, rivers, streams, and ponds.

This law is applicable to the building of logging roads.

Open-Air Fires

SECTION 2 – 12 – 6

A permit is required for open-air fires on or adjacent to forest land, under the authority of the Department of Environmental Management (DEM). Burning permits are issued by local fire departments, who act as DEM's authorized agents.

Registration of Forestry-Related Workers, Equipment, or Businesses

SECTION 2 – 15 – 1, 4

Persons cutting trees for commercial forest products must be registered annually. Owners cutting forest products for their own use, or cutting fewer than 5,000 board feet or 25 cords for sale to others, are exempt.

Threatened and Endangered Species

TITLE 20, SECTIONS 1 – 5

It is illegal to buy, sell, store, transport or otherwise traffic in any endangered plant or animal species. A permit is required for collection of endangered species of plants and animals for scientific or educational purposes. The Director of the Department of Environmental Management is responsible for listing of species and granting of permits.

Timber Trespass

SECTION 43 – 20 – 1

Any person who cuts, destroys, or carries away trees, timber, wood, or underwood from the land of another without permission must pay twice the value of any tree and thrice the value of wood or underwood.

Weights and Measures

SECTION 47 – 12 – 1

Cordwood is defined as any wood less than 4 feet in length. A cord is defined as 128 cubic feet of wood, closely packed.

VERMONT

VSA = Vermont Statutes Annotated

Current Use Assessment

USE VALUE APPRAISAL
TITLE 32, VSA 124, SECTIONS 3752 – 360

Eligibility requirements:

(1) A landowner must own at least 25 acres of contiguous forest land, exclusive of house site (2 acres surrounding any house). Owners of more acreage need not put their entire holding under classification.

(2) Open and nonproductive land cannot exceed 20 percent of the total parcel.

(3) A landowner must have and follow a fifteen-year management plan, which has the objective of growing forest crops, and is approved by the Department of Forest, Parks, and Recreation.

(4) Applications are due by September 1st.

Ongoing requirements:

(1) An annual report of compliance with the plan must be certified by the county forester.

(2) A fifteen-year plan need only be updated every five years. Conformance reports are required annually.

Recommended use values:

The state has suggested two use values for forest land, based upon the potential productivity of the soil.

Withdrawal:

If any owner subdivides the land into parcels smaller than 25 acres, constructs a building, or harvests contrary to the management plan, the land is considered to be "developed." The developed portion of the land is then taxed at 10 percent of its fair market value. Land can be withdrawn from the program at any time, but the land-use tax is not due until the land has been actually developed. The 10 percent tax is based on the fair market value at the time the land was withdrawn. If the state does not have sufficient funds to reimburse towns for the full amount of tax revenue lost as a result of use value appraisal, landowners who have received use value appraisal may pay the portion of their tax not reimbursed by the State and remain in the program. If they elect to withdraw from the program, they will not be liable for the land use change tax. Forest land does not become declassified when it is sold, unless its use changes.

TAX STABILIZATION AGREEMENTS
TITLE 24, VSA, SECTION 2741

By town meeting vote, Vermont towns may authorize their selectmen to enter tax stabilization contracts with owners of forest land to fix the amount of taxation on qualifying forest property. Both the qualifications and amount of tax relief are set by the town. Contracts may not exceed ten years duration, and must be available for public inspection.

TITLE 32, VSA, SECTION 3846

A town's board of selectmen, without voter approval, may enter tax stabilization contracts with qualifying forest land owners. While selectmen can determine the amount of tax relief to be granted, certain state requirements for property qualifications must be satisfied:

(1) Qualifying forest land must be at least 25 acres in size, and must be actively managed for repeated forest crops.

(2) Stabilization agreements must provide for a rollback tax, equivalent to the total amount of the tax relief granted to the landowner for the previous three years. This would be due if the land were converted to another use, in violation of the contract.

Liability of Landowners

TITLE 10, VSA, SECTION 5212(B)

If land is used for recreational purposes and no fee is charged, the owner shall not be liable for injuries unless injuries are a result of willful or reckless conduct by the owner.

TITLE 12, VSA, SECTION 5751

Agricultural activities, which include the commercial harvesting of trees, if carried out in a reasonable fashion consistent with good practices, are entitled to a "rebuttable presumption" to the effect that the activities do not constitute a nuisance. *Therefore, the burden of proof lies with the person making a nuisance complaint.*

Logging in Fragile or Specially Designated Areas

TITLE 10, VSA, SECTION 1021, 1025

A permit is required from the Agency of Environmental Conservation to change the course or current of any stream with a drainage greater than 10 square miles either by movement, fill, or excavation of 10 cubic yards of fill.

TITLE 10, VSA, SECTION 6001 (3), 6081

Any logging activity above 2,500 feet in elevation requires an Act 250 permit. *Act 250 is the law which regulates land development.*

TITLE 10, VSA, SECTION 259

No increase in the discharge of any wastes (soil, oil, or gas) which will degrade the quality of waters is permissable. Acceptable color and turbidity levels are stated in the law. If acceptable management practics (AMPs) are in place; that is, if loggers and landowners have followed proper measures to protect a waterway, no permits are necessary. The acceptable management practices have been established by the Department of Forest Parks and Recreation. A violation of this law occurs only if there is a discharge into a waterway and the AMPs have not been followed. "Slash" that includes branches, bark or pieces of wood in a stream are automatically considered a violation except in winter when used as a temporary bridge. Penalties could include restoration of the water quality, fines up to $25,000 and six months imprisonment.

TITLE 10, VSA, SECTION 1275 (C)

Wetland rules that affect forestry practices include the following: the configuration of the wetland's outlet cannot be altered, log landings in wetlands are restricted to frozen conditions, existing roads in wetlands used for silvicultural purposes cannot be increased in width by more than 20%, new road construction in wetlands requires approval from the Department of Environmental Conservation and acceptable management practices (AMPs) are mandatory in wetlands.

Open-Air Fires

TITLE 10, VSA, SECTION 1496

Burning permits are required for open-air fires, except when ground is snow-covered. Permits can be obtained from town forest-fire wardens.

Petition for Right-of-Way

TITLE 19, VSA, SECTIONS 325, 326

If it is necessary for the convenient removal of lumber or wood to pass through the lands of a person other than the owner of the lumber or wood, the selectmen, if petitioned, may lay out a right-of-way. Damages shall be assessed and paid before the right-of-way is used. The selectmen may set conditions for use.

Registration of Forestry-Related Workers, Equipment, or Businesses

TITLE 10, VSA, SECTION 2623 (3)

Whole-tree chip harvesters, portable sawmills, and other similar equipment must be licensed by the Department of Forest, Parks, and Recreation.

Slash Disposal

TITLE 10, VSA, SECTION 1259(A)

The discharge of any waste into any waters is not permitted.

TITLE 20, VSA, SECTION 2751

All slash must be removed from within 50 feet of the right-of-way of any public highway and of property boundary lines. All slash must be removed from within 100 feet of buildings on adjoining property.

Threatened and Endangered Species

TITLE 10, VSA, SECTIONS 5403 – 5409

It is illegal to take or possess any wild plant or wildlife listed by the state as threatened or endangered without a permit. Permits ("variances") may be granted by the Secretary of the Agency of Natural Resources who is also responsible for the listing of threatened and endangered species. The Secretary is advised by a nine-member committee of citizens and public officials. Agriculture and forestry activities are protected from "undue interference" and in most cases are exempt from permit requirements.

Timber Trespass

TITLE 13, VSA, SECTION 3606

Any persons who without permission willfully cut down, destroy, or carry away trees, timber, or wood from land of another shall be liable for treble damages. If persons acted by mistake or had good reason to believe the land was theirs, only single damages plus costs are required.

Weights and Measures

TITLE 9, VSA, SECTION 2693

Sawlogs or round timber shall be sold according to value obtained by the international quarter-inch log rule, or by Vermont Rule.

TITLE 9, VSA, SECTION 2651

The term "cord" is defined as the amount of wood that is contained in a space of 128 cubic feet when the wood is closely stacked.

The legal weight of a gallon of maple syrup is not less than 11.07 pounds, and the legal measure is 231 cubic inches (128 fluid ounces) at 68 degrees F.

Sources of Information and Assistance

An updated version of this appendix is available at
www.upne.com/appendices/woodland.html

This appendix contains a state-by-state list of organizations (public and private) and of directories which provide information and assistance to New England woodland owners. For each state, sources are listed in the following order:

> State and federal agencies
> Woodland owners organizations
> Directories
> Market reports (stumpage prices)

The main offices of state and federal agencies that are listed here will provide information for locating their local offices. For woodland owners organizations, only permanent addresses and telephone numbers are included; to obtain current information about any of the other woodland owners organizations that are listed, an owner may contact a local county forester (called a service forester in some New England states) or Extension office. Directories listed can be obtained from a local county forester or Extension office, unless otherwise noted.

CONNECTICUT

State and Federal Agencies

Service Foresters

Dept. of Environmental Protection
209 Hebron Ave.
Marlborough, CT 06447
(203) 295-9523

Dept. of Environmental Protection
Goodwin Conservation Center
23 Potter Road
No. Windham, CT 06256-1616
(203) 455-0699

Dept. of Environmental Protection
State Forest Tree Nursery

RFD #1, Box 23A
Voluntown, CT 06384
(203) 376-2513
Dept. of Environmental Protection
Valley Headquarters
P.O. Box 161
Pleasant Valley, CT 06063
(203) 379-9085

Main Offices

Agricultural Stabilization and Conservation Service
88 Day Hill Road
Windsor, CT 06095
(203) 285-8483
Department of Environmental Protection, Forestry Unit
State Office Building
165 Capitol Ave.
Hartford, CT 06106
(203) 566-5348
Department of Environmental Protection, Wildlife Unit
State Office Building, Rm. 254
165 Capitol Ave.
Hartford, CT 06106
(203) 566-4683
Cooperative Extension Service Forester
Department of Natural Resources
University of Connecticut
Storrs, CT 06268
(203) 486-2840
USDA Soil Conservation Service
16 Professional Park Road
Storrs, CT 06268
(203) 487-4013

Woodland Owners Organizations

Connecticut Forests and Parks Association
16 Meriden Road
Rockfall, CT 06481-2961
(203) 346-2372
Eastern Connecticut Forest Landowners Association
P.O. Box 404
Brooklyn, CT 06234
Connecticut Christmas Tree Growers Association

Connecticut Tree Farm Committee

Maple Syrup Producers Association of Connecticut

Directories

Primary Wood Processors

Eastern CT Forest Landowners Annual Directory of Local Resources for Forest Owners

DEP Natural Resources Information Directory & List of Publications

DEP Connecticut Directory of Environmental Organizations

Market Reports

Southern New England Annual Forest Products Report

Quarterly Stumpage Reports

MAINE

State and Federal Agencies

County foresters are usually listed in the yellow pages of the telephone book under the headings "Government Offices–State; Conservation Department; or Regional Forester; or their numbers can be obtained by contacting the head office of the Bureau of Forestry, which is listed below.

Main offices

Agricultural Stabilization and Conservation Service
P.O. Box 406
Bangor, ME 04402-0406
(207) 942-0342

Extension Forester
107 Nutting Hall, University of Maine, Orono, ME 04473
(207) 581-2890 or (800) 287-0274 (in-state)

Maine Bureau of Forestry
State House Station 22, Augusta, ME 04333
(207) 289-2791 or (800) 367-0223 (in-state)

Maine Dept. of Inland Fisheries and Wildlife
284 State St., Augusta, ME 04333
(207) 289-2871

USDA Soil Conservation Service
5 Godfrey Drive, Orono, ME 04473
(207) 866-7241

Woodland Owners Organizations

Maine Christmas Tree Growers Association

Maine Forest Products Council
146 State St., Augusta, ME 04330
(207) 622-9288

Maine Maple Producers

Maine Tree Farm Committee

Small Woodland Owners Association of Maine
P.O. Box 926, Augusta, ME 04332
(207) 626-0005

Directories

Loggers (Bureau of Forestry)

Primary Processors (Bureau of Forestry)

Directory of Forestry Services (Bureau of Forestry)

Yankee Woodlot Directory (State Extension Forester)

Market Reports

Stumpage and mill-delivered price list (Bureau of Forestry)

MASSACHUSETTS

State and Federal Agencies

Service foresters in counties and regions

DEM Division of Forests and Parks
P.O. Box 1433, Pittsfield, MA 01202–1433
(413) 442–8928

DEM Division of Forests and Parks
P.O. Box 484, Amherst, MA 01004–0484
(413) 545–5993

DEM Division of Forests and Parks
P.O. Box 155, Clinton, MA 01510-0155
(508) 368-0126

DEM Division of Forests and Parks
P.O. Box 829, Carlisle, MA 01741-0829
(508) 369-3350

DEM Division of Forests and Parks
P.O. Box 66, South Carver, MA 02366-0066
(508) 866-2580

Main offices

Agricultural Stabilization and Conservation Service
445 West St., Amherst, MA 01002
(413) 256-0232

Dept. of Environmental Management
Division of Forests and Parks
100 Cambridge St., 19th Fl., Boston, MA 02202
(617) 727-3180
Extension Forester
Holdsworth Natural Resources Center
University of Massachusetts, Amherst, MA 01003
(413) 545-2665
Dept. of Fisheries, Wildlife & Env. Law Enforcement
Division of Fisheries and Wildlife
100 Cambridge St., 19th Fl., Boston, MA 02202
(617) 727-3151
USDA Soil Conservation Service
441 West St., Amherst, MA 01002
(413) 256-0441

Woodland Owners Organizations

Massachusetts Forestry Association
P.O. Box 1096, Belchertown, MA 01007-1096
(413) 323-7326
Massachusetts Christmas Tree Growers Association
Massachusetts Maple Producers Association
Massachusetts Tree Farm Committee

Directories

Commercial Loggers, Sawmills and Dry Kilns in Massachusetts (DEM)
Forestry Services in Massachusetts (Extension)
Massachusetts Natural Resource Agency Directory (Extension)
Practicing Foresters (DEM)

Market Reports

Southern New England Annual Forest Products Report
Quarterly Stumpage Reports

NEW HAMPSHIRE

State and Federal Agencies

Local Extension Foresters, who are the same as service or county foresters, are listed in the white and yellow pages of the telephone book under "County Government; Cooperative Extension"; or "University of New Hampshire–Cooperative Extension."

Main offices

Agricultural Stabilization and Conservation Service
P.O. Box 1398, Concord, NH 03302-1398
(603) 244-7941

Department of Resources and Economics, Division of Forests and Lands
P.O. Box 856, Concord, NH 03302-0856
(603) 271-2411

Chief Extension Forester
Rm. 108, Pettee Hall, University of New Hampshire
Durham, NH 03824
(603) 862-1028

New Hampshire Fish and Game Department
2 Hazen Drive, Concord, NH 03301
(603) 271-3421

USDA Soil Conservation Service
2 Madbury Road, Federal Building, Durham, NH 03824
(603) 868-7581

Woodland Owners Organizations

New Hampshire Timberland Owners Association
54 Portsmouth St., Concord, NH 03301
(603) 224-9699

Society for the Protection of New Hampshire Forests
54 Portsmouth St., Concord, NH 03301
(603) 224-9945

Directories

List of Consulting Foresters
Directory of Sawmills
Owning a Piece of the Forest (sources of information and services)

Market Reports

New Hampshire Forest Market Report

RHODE ISLAND

State and Federal Agencies

Service foresters

County Forester
2185 Putnam Pike, West Gloucester, RI 02814
(401) 568-2013

County Forester
Arcadia Mgt. Area, 260 Arcadia Rd., Hope Valley, RI 02832
(401) 539-2356

Main offices

Agricultural Stabilization and Conservation Service
60 Quaker Lane, Rm. 40, Warwick, RI 02893
(401) 828-8232

Division of Forest Environment
1037 Hartford Pike, North Scituate, RI 02857
(401) 647-3367

Extension Specialist, Forestry and Wildlife Management
University of Rhode Island, Kingston, RI 02881
(401) 792-2370

USDA Soil Conservation Service
60 Quaker Lane, Rm. 46, Warwick, RI 02893
(401) 828-1300

Woodland Owners Organizations

Rhode Island Forest Conservators Organization
Rhode Island Christmas Tree Growers Association
Rhode Island Tree Farm Committee

Directories

Consulting Foresters List
Woods Operators
Sawmills

Market Reports

Southern New England Annual Forest Products Report
Quarterly Stumpage Report

VERMONT

State and Federal Agencies

County foresters and local soil conservation scientists are listed at the front of the telephone book on the page giving community service numbers under the heading Environment. Local Agricultural Stabilization and Conservation Services are listed in the yellow pages under Government Offices–United States Department of Agriculture.

Main offices

Agricultural Stabilization and Conservation Service
346 Shelburne St., Burlington, VT 05401
(802) 951-6715

Extension Forester, University of Vermont
Natural Resource Office, Aiken Center, Burlington, VT 05405
(802) 656-2913

Vermont Department of Forests, Parks and Recreation
103 South Main St., Waterbury, VT 05671-0601
(802) 244-8715

Vermont Fish and Game Department
103 South Main St., Waterbury, VT 05671-0601
(802) 244-7331

USDA Soil Conservation Service
P.O. Box 69, Winooski, VT 05404
(802) 951-6795

Woodland Owners Organizations

Vermont Land Trust
8 Bailey Ave., Montpelier, VT 05602
(802) 223-5234 or (800) 639-1709

Vermont Timberland Owners Association

Woodland Owners Association

Vermont Tree Farm Committee

Directories

Consulting Foresters
Forest Management Information and Services
Vermont Sawmill Operators

Market Reports

Vermont Forest Quarterly

Index

Library of Congress Cataloging-in-Publication Data

Beattie, Mollie, 1947–1996.
    Working with your woodland : a landowner's guide / Mollie Beattie,
Charles Thompson, and Lynn Levine; illustrations by Nancy Howe.
— Rev. ed.
      p.   cm.
    Includes bibliographical references and index.
    ISBN 0–87451–622–6
    1. Forest management—New England. I. Thompson, Charles, 1947– .
II. Levine, Lynn, 1952– . III. Title.
SD144.A12B4   1993
634.9'0974—dc20   92–56900
∞